Aerodynamics

Edited by Mofid Gorji-Bandpy and Aly-Mousaad Aly

Published in London, United Kingdom

IntechOpen

Supporting open minds since 2005

Aerodynamics
http://dx.doi.org/10.5772/intechopen.78935
Edited by Mofid Gorji-Bandpy and Aly-Mousaad Aly

Contributors
Sadiq Raji, Patricia Popoola, Sisa Pityana, Olawale Popoola, Fatai Aramide, Modupeola Dada, Samson Adeosun, Ntombi Mathe, Hoang Thi Kim Dung, Nguyen Phu Khanh, Anna Markhotok, Vladimir Aleksievich Frolov, Chedhli Hafien, Pedro Junior Zucatelli, Davidson Moreira, Erick Giovani Sperandio Nascimento, Alex Santos, Alejandro Arce, Aly-Mousaad Aly, Sergey Gandzha, Anton Kotov, Aram Baghiyan, Faiaz Khaled, Hamzeh Gol-Zaroudi, Adnen Bourehla, Mounir Bouzaiane, Monnamme Tlotleng, Nana Kwamina Kum Arthur

First published in London, United Kingdom, 2021 by IntechOpen
IntechOpen is the global imprint of INTECHOPEN LIMITED, registered in England and Wales, registration number: 11086078, 5 Princes Gate Court, London, SW7 2QJ, United Kingdom
Printed in Croatia

British Library Cataloguing-in-Publication Data
A catalogue record for this book is available from the British Library

Additional hard and PDF copies can be obtained from orders@intechopen.com

Aerodynamics
Edited by Mofid Gorji-Bandpy and Aly-Mousaad Aly
p. cm.
Print ISBN 978-1-83880-167-0
Online ISBN 978-1-83880-168-7
eBook (PDF) ISBN 978-1-83880-003-1

For EU product safety concerns:
IN TECH d.o.o., Prolaz Marije Krucifikse Kozulić 3, 51000 Rijeka, Croatia, info@intechopen.com or visit our website at intechopen.com.

Meet the editors

Dr. Mofid Gorji-Bandpy received his MS in Mechanical Engineering from the Faculty of Engineering, University of Tehran, Iran, in 1978. In 1990, he obtained his Ph.D. in Hydraulic Engineering from the School of Engineering, University of Wales College of Cardiff (UWCC), UK. At present, he is a full professor in the Department of Mechanical Engineering at the Babol Noshirvani University of Technology, Babol, Iran. He is also a visiting professor in the Department of Mechanical and Industrial Engineering at the University of Toronto, Canada. His major interests are in advanced methods of energy-conversion systems, turbo machinery, fluid mechanics, water distribution networks, and solutions to both energy and environmental problems. He has published more than 300 papers in these fields. He has also published several refereed publications in different fields of mechanical engineering, applied mathematics, and aerodynamics. In addition, Dr. Gorji-Bandpy was recognized as a Highly Cited Researcher by Clarivate Analytics in 2017 and 2018.

Dr. Aly-Mousaad Aly is an associate professor at Louisiana State University (LSU). His research aims at creating fundamental knowledge in wind engineering, dynamics, and control to build more resilient, economic, and recoverable infrastructure to enhance safety and reduce the huge cost of rebuilding after windstorms and earthquakes. He is the director of the LSU Windstorm Impact, Science, and Engineering (WISE) research and education program (wise.lsu.edu). His research enables advanced wind testing, which yields loads that mimic those at full scale. His research advanced the implementation of smart dampers in super-tall structures and a novel probabilistic theory that enables the analytical modeling of semi-active damping. He received a Ph.D. from the Polytechnic University of Milan, and an MSc and BSc from Alexandria University, Egypt.

Contents

Preface

Aerodynamics is a branch of physics concerned with the motion of air and the resulting forces acting on solid objects. In other words, aerodynamics is primarily related to the forces of drag and lift, and moments on airfoils, wings, fuselages, engine nacelles, and most importantly, whole airplane configurations, which are caused by air passing over and around solid bodies. It is also concerned with the design of automobiles, high-speed trains, and ships, as well as with the construction of such structures as bridges and tall buildings to determine their resistance to high winds. The basic principles are relatively easy to understand; they include how fluids flow in different ways. Most of the equations from fluid dynamics apply to aerodynamics as well, including all the governing equations, turbulence, boundary layer theory, and ideal gas assumption. This book presents novel ideas that traverse traditional disciplinary boundaries as well as introduces a wide spectrum of viewpoints and approaches in applied aerodynamics of the third millennium. The book will be of interest to those working in the fields of propulsion systems, power generation systems, the flight of aircraft, fluid flow, combustion, and other phenomena.

Mofid Gorji-Bandpy
Babol Noshirvani University of Technology,
Iran

Aly-Mousaad Aly
Louisiana State University,
United States of America

Section 1

Aeroelasticity

Chapter 1

Application of an Asynchronous Synchronous Alternator for Wind Power Plant of Low, Medium and High Power

Sergey Gandzha and Anton Kotov

Abstract

The chapter shows the prospects for the development of alternative energy. The growth rate of wind energy is ahead of other types of alternative energy, but the development of wind energy is constrained by a number of technical contradictions that need to be solved. The main problem of wind power is that the power and direction of the wind flow are continuously changing. This leads to the fact that the frequency of rotation of the generator constantly varies and alternator produces energy with nonstandard parameters in amplitude, frequency and phase. Converting this energy into energy with standard parameters is a difficult technical task. A brief analysis of the different directions to solve this problem is shown. It is proved that the promising direction of solving this problem from the point of view of efficiency is the use of double-fed induction alternator (DFIA). The chapter describes the principle of operation of the DFIA, the theory of energy conversion based on equivalent circuits. Approaches to the optimal design of generators based on generalized variables are shown. Two variants of the generator design are described. One option is to contain an additional exciter generator to power the rotor. The design of this 10 kW generator is presented. In another version, the power supply function of the rotor is performed by the battery. In addition, the battery performs the function of accumulation of electricity. It is concluded that the development of wind power in the direction of the DFIA is promising. On the basis of the proposed concept, a number of wind power plants can be built with power from 10 kW to 6 MW. DFIA can operate in standalone mode and in conjunction with electrical grid. The design for the range of wind turbines will be the same type. The DFIA will differ only in size.

Keywords: windmill, double-fed induction alternator, asynchronous synchronous alternator, electronic control system, brushless electric machine, exiting alternator, standard energy, nonstandard energy

1. Introduction

The last 10 years in the world scientific community there are heated discussions about the ways of development of modern energy. The conservative part of scientists believes that hydrocarbon energy based on the use of coal, oil and gas will remain the main source of heat and electricity for economically developed countries for a long time. There are grounds for this view. Hydrocarbon reserves on earth are still quite large. The hydrocarbon industry has a well-established infrastructure. There are enterprises for the production of the necessary equipment, enterprises for the construction and maintenance of energy facilities, scientific results to improve the efficiency of electricity generation in this way.

The advantages of the traditional energy system based on hydrocarbons, is faced with many problems and contradictions. These contradictions are growing every year and, ultimately, can lead to a crisis of traditional energy, if you do not attempt to develop alternative energy.

One of the reasons is the rise in the cost of hydrocarbon production. Cheap oil, gas and coal have already have used. We have to penetrate to great depths under the earth's surface, to work at the bottom of the ocean, to explore the remote Northern areas for mining. It may not be economically feasible after all.

The second very important reason is environmental problems. Nature has been creating an ecological balance of carbon dioxide in the atmosphere for millions of years. Hydrocarbon power industry disturbs this balance. Humanity has not yet fully appreciated the consequences of this imbalance. The crisis can affect more than just climate change. The next step is a violation in nature. At the final stage, problems will arise in the social community. But, probably, it will be too late to solve these problems.

Very little is said about the low efficiency of traditional energy. The highest efficiency of thermal power plants is about 30%. According to various estimates, their share in the global energy balance ranges from 60 to 70%. These figures show that we heat our planet and clog it with emissions of production. Nature will inevitably take revenge on humanity for these actions.

The development of technical systems has its own laws with its stages of development [1]. Qualitatively, this process is shown in **Figure 1**.

At the first stage, the technical system is born. This is the stage of laboratory research and scientific analysis. This stage requires a lot of effort from scientists. As a rule, this stage requires large financial investments in scientific research and is not commercial.

At the second stage, the technical system is developing very quickly. Industrial production shows great interest in the system, because it gives a commercial result.

At the third stage, the technical system comes to the saturation phase. Staying at a high level, it has exhausted its resources. The pace of its development is very slow. Further investment of finances in its development does not give a commercial result.

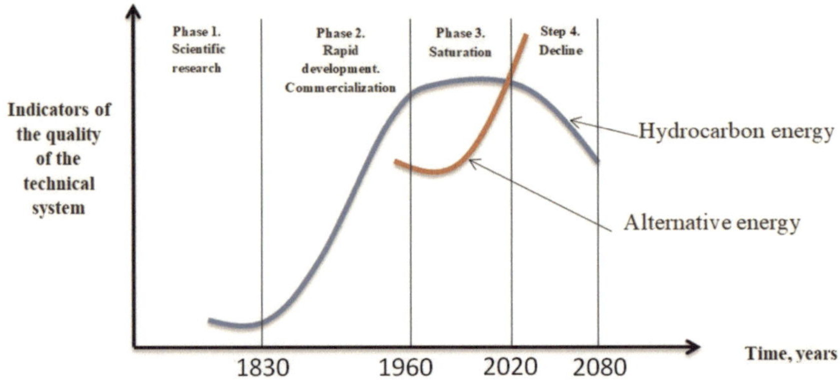

Figure 1.
Stages of development of technical systems.

At the fourth last stage of its development, the technical system is in decline. Industry is losing interest in its reproduction. It becomes unprofitable.

Hydrocarbon energy has more than 150 years of its development. According to various experts, it finishes the second phase of its development and moves into the third phase. The third phase can last for several decades and will inevitably come the phase of decline.

The theory of development of technical systems determines that in this process at a certain stage of development of the first system, the second alternative technical system is born. At first, it is difficult to compete with the dynamically developing first system, but it goes through all 4 phases of development and replaces the first technical system when it comes to decline.

This function is currently performed by alternative energy. **Figure 2** shows this tendency.

The question arises in what direction and how to develop alternative energy.

Alternative energy includes different types:

- Geo thermal energy;
- Water power energy;
- Bio gas energy;
- Solar energy;
- Wind energy.

The diagram shows that the share of wind energy in the total balance is the largest today.

Among others renewables the windmills are the most developing devices due to several advantages:

- Wind is the endless power source.
- It is an ecologically clean device.

Wind energy trends should be taken into account. It is shown in **Figure 3**.

It should be noted that the wind turbines are really environmentally friendly devices. They do not produce harmful emissions during their work and do not require dirty technology for their manufacture. In the production of solar panels are used a harmful production. Thus, the development of wind energy is an urgent task.

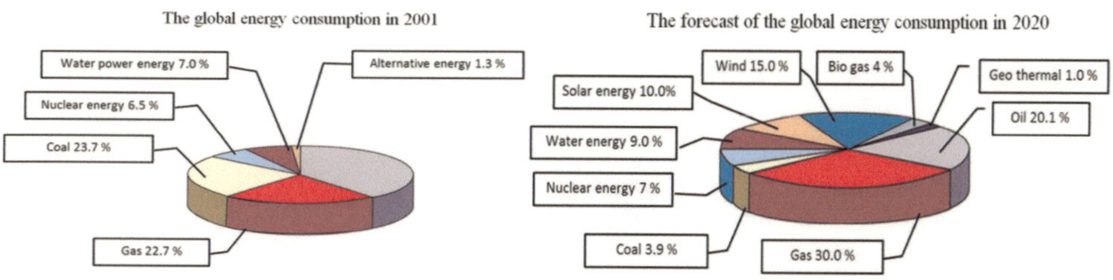

Figure 2.
Forecast of alternative energy development.

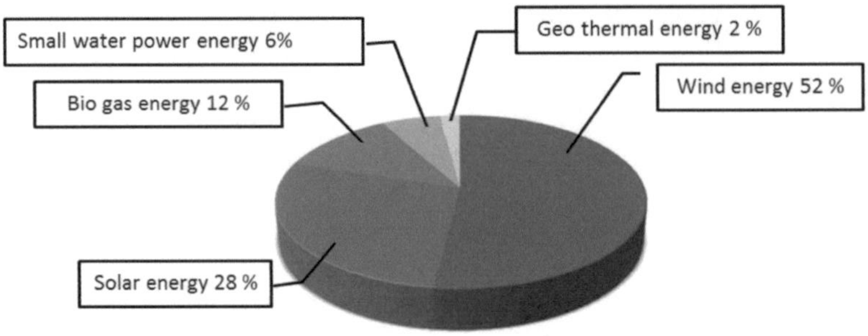

Figure 3.
Share of wind energy in the total balance of alternative energy.

2. Various of the designs of wind turbines

Wind turbines can be classified into the following types by output power

- wind turbines of low power (up to 10 kW);
- wind turbines of average power (from 10 to 100 kW);

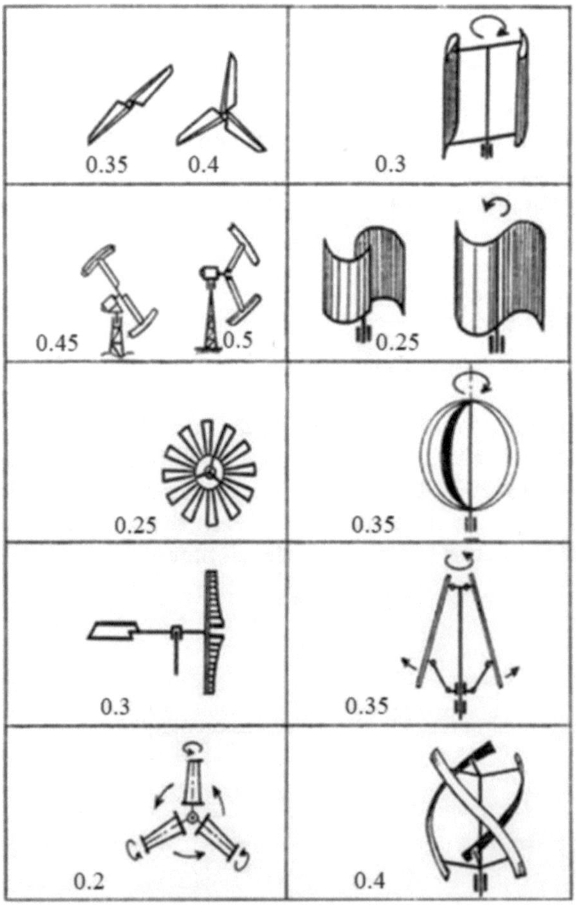

Figure 4.
Different designs of wind wheels and their efficiency.

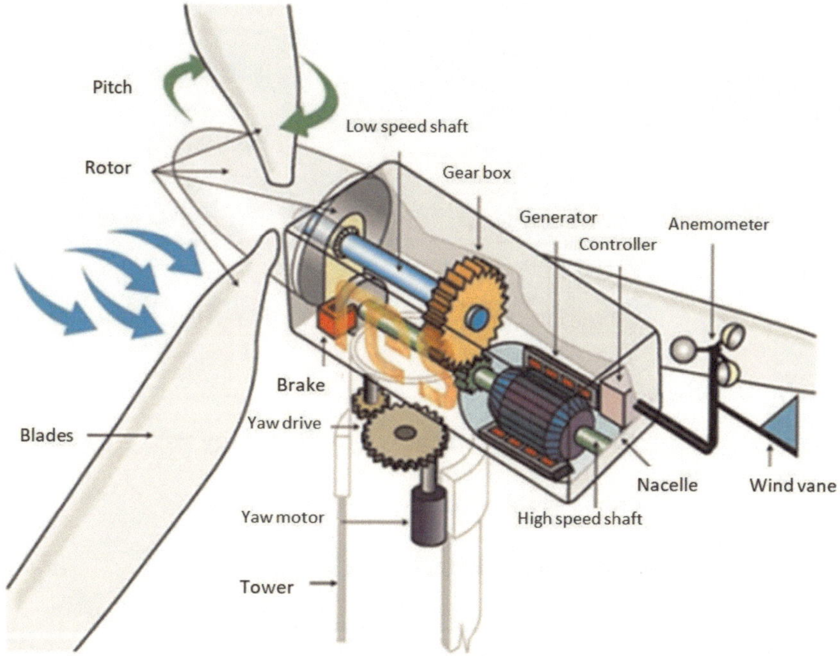

Figure 5.
The main elements of the wind turbine with a horizontal axis of rotation.

- high power wind turbines (more than 100 kW).

Wind power plants have many versions, but they can be divided into two main types

- Wind turbines with horizontal axis of rotation;

- Wind turbines in the vertical axis of rotation;

The efficiency of the wind wheel of a wind turbine is defined as the ratio of the mechanical power on the shaft to the power of the wind flow. It depends on the wind speed and the design of the wind wheel. The maximum efficiency for different wind turbines is shown in **Figure 4**.

Wind turbine is a complex technical device. It consists of many components and parts [2–4]. The design of the wind turbine with a horizontal axis of rotation is shown in **Figure 5**.

The most important elements of the wind turbine are an electric alternator and an electronic control system. The efficiency of the entire wind turbine depends on the correct choice of the alternator type.

3. The main problem of wind power

It is necessary to solve very serious problem for using windmills effectively [5–7].

The main problem of windmill's energy is the changing of the speed and the direction of wind flow. Thus, the windmill rotor rotates with a different speed and alternator gives the voltage with variable parameters of amplitude, frequency and phase (**Figure 6**). It is impossible to use this nonstandard energy. It is

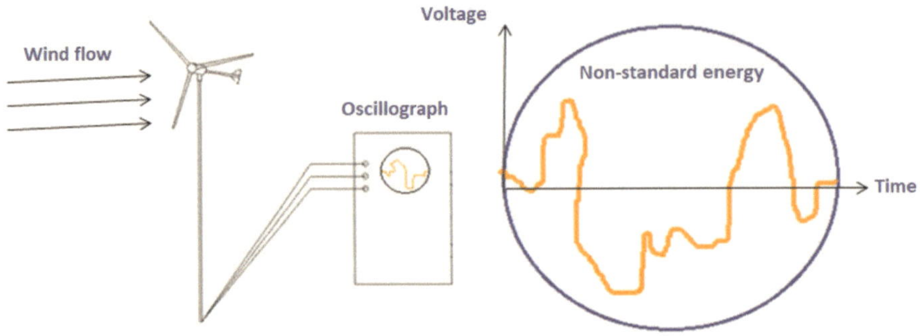

Figure 6.
Illustration of wind turbine operation.

necessary to transform the nonstandard energy to the energy with standard parameters [4, 8–18].

In practice some methods to solve this problem are known.

1. Application of asynchronous (induction) alternator.

Figure 7 shows the design of the inductor alternator.
The disadvantages of this method are:

- It is necessary to supply the reactive energy for working of the induction alternator. Such windmills can work only with electric grid or with battery of capacitors;

- When induction alternator works with electric grid it uses the reactive power and makes parameters worse.

2. Application of permanent magnet synchronous alternator with rectifier and invertor.

Figure 8 shows the design of synchronous alternator.
The disadvantages of this method are:

- They are many stages for transforming the nonstandard energy to standard energy (the rectification, the stabilization, the inversion). Every stage has his own loses and efficiency. So, the total efficiency is very low.

- The equipment for this method is very expensive.

3. Application of an asynchronous synchronous alternator (double fed inductor alternator) (DFIA)

Figure 7.
Design of the inductor alternator.

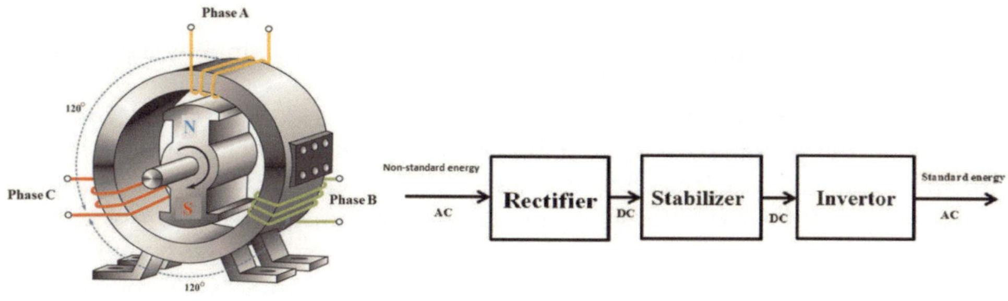

Figure 8.
Design of the synchronous alternator.

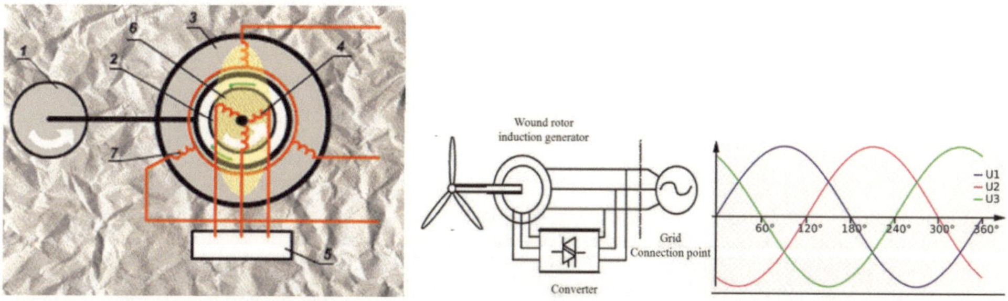

Figure 9.
Design of the asynchronous synchronous alternator (1, wind turbine; 2, rotor; 3, Stator; 4, rotor; 3 phases winding; 5, electronic control system; 6, rotating magnetic field; 7, Stator 3 phases winding).

Figure 9 shows the design of the double fed alternator.

This type of alternator can transform the nonstandard energy into standard energy without complex units and without additional stages of energy conversion. This is a new approach to solving the problems of wind energy. This Chapter will be devoted to the description of this promising wind turbine.

4. The application of the DFIA

It is well-known the principle of operation of the double-fed alternator [12, 14–16]. Electric machine of this type has been studied by scientists around the world since 1960. It has different applications, but in wind power it is used only for the last 10 years. The authors offer a unique design that they have developed for a row of wind power plants. The design has a Russian patent. It is described below.

The alternator consists of two electric machines. The main electric machine is a power alternator. It is the induction machine with a phase rotor. So, the stator and the rotor have three phase windings. The alternator has the inverse design. It means that the rotor is settled outside of stator. It is very comfortable for the windmill because it is possible to attach the blades to the rotor. The three phase rotor winding is supplied by an adjustable voltage source. The rotor creates the rotating magnet field. This magnet field generates the voltage in the stator winding. If the rotor is fixed the speed of rotation of magnet field depends only from frequency of voltage source. But if the rotor begins to rotate, the speed of magnet field is added to the speed of rotor rotation. So, the total speed of magnet field is combined with two parts. We can change the frequency of the adjustable three phase voltage sources. When the speed of rotor is variable we can choose the necessary frequency of voltage source and we will have the stable frequency in the stator winding. In this way, we solve the problem of voltage frequency stabilization with different speed of rotor rotation. The amplitude of stator voltage is regulated by amplitude of rotor winding voltage.

But it is necessary to solve one serious problem to provide the reliability to windmill. It is impossible to use the sliding contact for windmill because it is very difficult to serve this device at height and in a difficult environment. The system has the second alternator (exciter) for brushless current supply. The brushless electric machine with combine excitation is used for the exciter. This type of electric machine has the stable external voltage with different speed of rotation. It is very comfortable in this case. In details, this type of electric machine is described in [1, 4, 19].

The sketch of alternator with exciter, 3D model and diagram of electronic control system is present in **Figures 10–12**. They allow understanding the principle of operation more carefully.

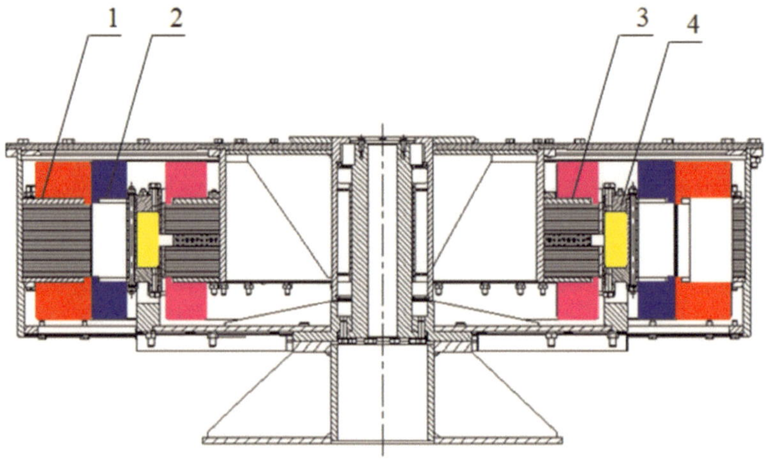

Figure 10.
Sketch of double-fed alternator with exciter. (1, rotor of power alternator; 2, stator of power alternator; 3, rotor of excitation alternator; 4, stator of exciter alternator).

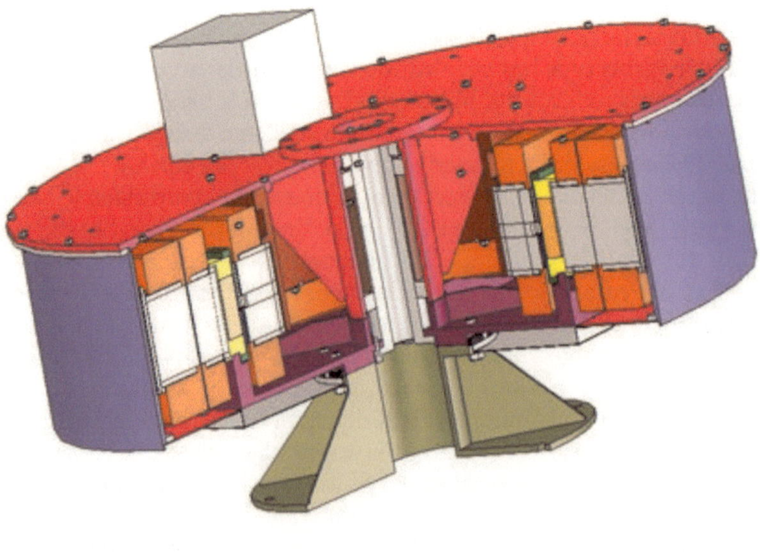

Figure 11.
3D model of the windmill alternator.

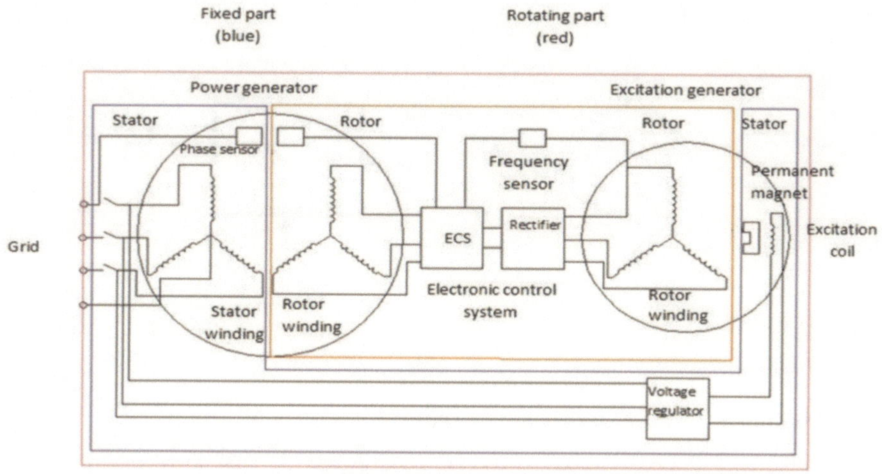

Figure 12.
Diagram of electronic control system.

The electronic control system is built in to the alternators. This concept allows having very high efficiency and very low cost.

This idea was realized for the windmill 10 kW but the concept can be scaled to megawatts sizes.

5. Theory of double-feed induction machine

The beginning of the work in the field of the theory of DFIA was initiated by the research in the late 1960s — early 1970s by the laboratory of double feeding machines under the guidance of doctors of technical sciences M. Botvinnik and Yu. G. Shakaryan [20].

The results of these works were the development of the foundations of the theory, structural schemes, principles of construction, as well as the creation of a number of experimental and industrial installations.

In a double-feed machine, energy is supplied (or withdrawn) to both the stator and rotor windings. This allows you to control the flow of active and reactive power of the dual-feed machine. The expediency of using DFIA is observed in the construction of alternator sets with a variable speed of rotation of the shaft, which is very important for wind power plants.

The stator windings of the machine are connected to the power supply network (U_1 = const, f_1 = const), and the rotor windings are connected to a separate controlled power source (active semiconductor converter ASC).

There are various options for implementing a controlled power supply in the rotor circuit. A scheme of independent rotor power supply is interested in wind power plants (**Figure 13**).

The active Converter is a three-phase bridge circuit of the inverter, made on transistors. The main property of the active Converter is the ability to two-way transmission of active power. When transmitting power from an AC source, ASC acts as an active controlled rectifier (ACR). When transmitting power from the DC link, ASC acts as an Autonomous Inverter (AI). Such structures allow you to control the flow of active and reactive power between the supply network and the machine.

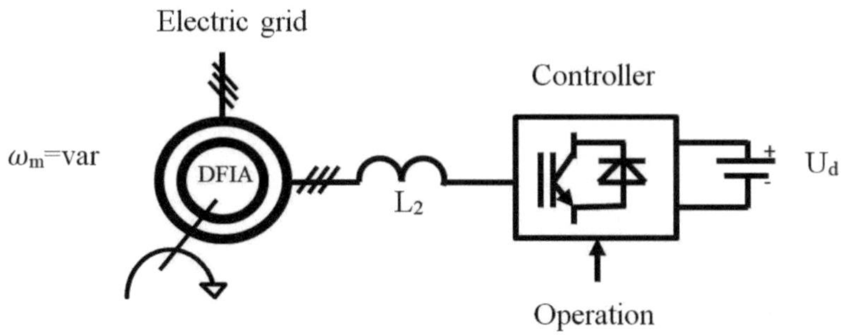

Figure 13.
Controlling of rotor power.

The main advantages of DFIA are as follows:

1. Control is carried out on the rotor circuit; this makes the high energy efficiency of the system.

2. DFIA can operate at speeds above and below synchronous, both in motor mode and in alternator mode. Therefore, when using DFIA as a alternator, there are no difficulties with stabilizing the parameters of the generated voltage when changing the speed of the shaft rotation, which is important for wind turbines.

Modes of operation, vector and energy diagrams.
Theoretically, the following operating modes of DFIA are possible:

- motor mode at a speed below synchronous;
- motor mode at a speed above synchronous;
- alternator mode at a speed higher than synchronous;
- alternator mode at a speed below synchronous.

In all these modes, the system of equations describing the electromagnetic processes of the machine in a steady state using the method of the resulting vector [21], has the form:

$$\overline{U}_1 = R_s\overline{I}_1 + jX_{ls}\overline{I}_1 + \overline{E}_1 \qquad \frac{\overline{U}_2^{|}}{s} = \frac{R_r}{s}\overline{I}_2^{|} + jX_{lr}\overline{I}_2^{|} + \overline{E}_2 \qquad \overline{I}_m = \overline{I}_1 + \overline{I}_2^{|}, \tag{1}$$

where $\overline{U}_1$ is the line voltage; $\overline{I}_1, \overline{E}_1 = j\omega_1\overline{\Psi}_m = jX_m\overline{I}_m$ is the current and EMF in anchor winding; $\overline{U}_2^{|}, \overline{I}_2^{|}, \overline{E}_2^{|} = j\omega_1\overline{\Psi}_m = jX_m\overline{I}_m$ is the led voltage, current and EMF in rotor winding; $\overline{I}_m$ magnetizing current; $R_s, X_{ls} = \omega_1 L_{ls}$ is the active and inductive resistance of the stator; $R_r, X_{lr} = \omega_1 L_{lr}$ is the led active and inductive resistance of the rotor; $X_m = \omega_1 L_m$ is the inductive reactance of magnetization circuit; s is the slide. The T-shaped equivalent circuit based on Eq. (1) is shown in **Figure 14a**.

Let us transform the T-shaped equivalent circuit into a G-shaped equivalent circuit. Conversion factor is:

$$C = 1 + \frac{X_{ls}}{X_m}. \tag{2}$$

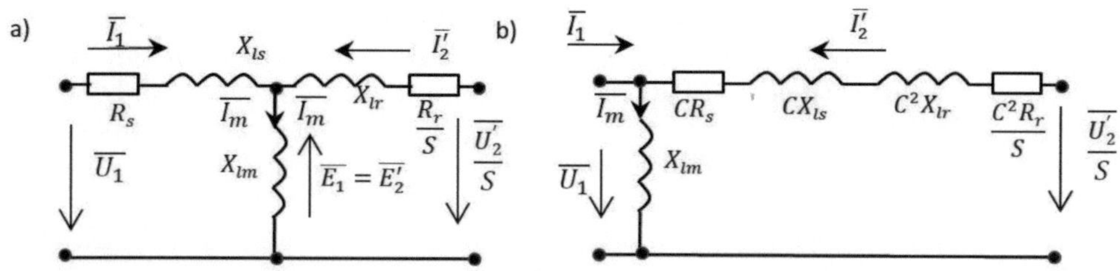

Figure 14.
Equivalent circuit of the DFIA: a) T-shaped equivalent circuit; b) G-shaped equivalent circuit.

We write down the equations for the G-shaped equivalent circuit (3)

$$\overline{U}_2^{|} = \overline{U}_1 + R_2\overline{I}_2^{|} + jX_2\overline{I}_2^{|};\ \overline{U}_1 = jX_{1m}\overline{I}_m, \overline{I}_m = \overline{I}_1 + \overline{I}_2^{|};\ R_2 = R_sC + \frac{R_r}{s}C^2 \approx R_s + \frac{R_r}{s};$$
$$X_2 = X_{ls}C + X_{lr}C^2 \approx X_{ls} + X_{lr};\ X_{1m} = X_{ls} + X_m \tag{3}$$

The following should be taken into account when assessing energy flows

1. The active power in the source is positive when the source gives energy and negative when the source consumes energy.

2. The reactive power in the source is positive (inductive) when the voltage is ahead of the current, and negative (capacitive) when the voltage is behind the current.

3. The mechanical power on the shaft is positive in the motor mode of your operation and negative in the generator mode of your operation. Active and reactive power sources $\overline{U}_1$ and $\overline{U}_2^{|}$ are determined from the expressions.

$$P_1 = \frac{3}{2}U_1I_1\ cos\ \varphi_1;\ P_2 = \frac{3}{2}\overline{U}_2^{|}\overline{I}_2^{|}\ cos\ \varphi_2;\ Q_1 = \frac{3}{2}U_1I_1\ sin\ \varphi_1;\ Q = \frac{3}{2}\overline{U}_2^{|}\overline{I}_2^{|}\ sin\ \varphi_2 \tag{4}$$

In the direction of the mechanical power acting on the shaft, the modes can be divided into motor mode and generator mode.

We consider only all generator modes for a wind turbine.

The mechanical power reported to the shaft of the machine at n < n0 (**Figure 15**) and the electrical power obtained by the rotor through a semiconductor converter, are transmitted to the stator and after deduction of losses are recovered to the grid.

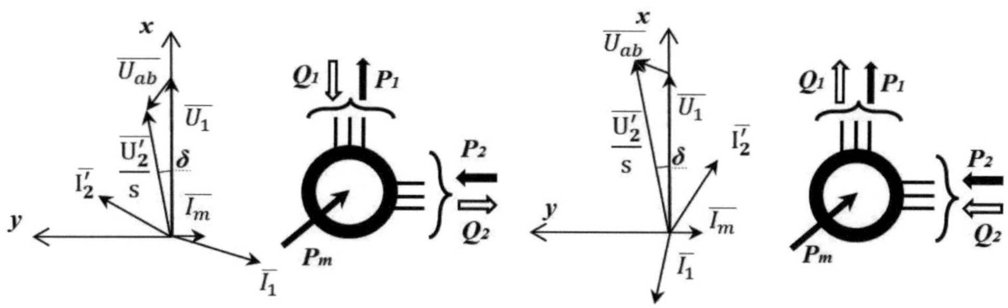

Figure 15.
Vector diagram and energy diagram of DFD in generator mode at $n < n_o$.

Energy diagrams of the generator mode DFIA at n > n0 are shown in **Figure 16**. In this mode, the mechanical power supplied to the shaft from the rotor side is converted into electrical power. Part of this power is given to the network from the rotor side. The other part is given to the stator circuit in the form of electromagnetic power and after deduction of losses is recovered in the network.

The balances of power in the various modes are summarized in **Table 1**.

A system of equations should be solved to quantify the properties of DFIA. This solution is carried out in a rotating coordinate system x (real axis) and y (imaginary axis). In this case, the voltage U1 is combined with the real axis x, and for the secondary voltage the ratio $U_{2m} = \frac{\omega 1U2}{\omega 2}$ and the phase σ are given. It is important to emphasize that when the real axis is combined with the voltage vector, the real component of the current is the active current of the stator, and the imaginary component is the reactive current of the stator.

$$I1 = \frac{\overline{(U1|)}\left(\frac{Rr}{s} + jXs\right) - j\frac{\overline{U2|}}{s}Xm}{(Rs + jXs)\left(\frac{Rr}{s} + jXs\right) - X_m^2}; \quad \overline{I2} = \frac{\frac{\overline{U2|}}{s}(Rs + jX) - j\overline{U1|}\,Xm}{(Rs + jXs)\left(\frac{Rr}{s} + jXs\right) - X_m^2}. \tag{5}$$

The total current of the rotor and stator are calculated by the formula:

$$I1 = \sqrt{I_{1x}^2 + I_{1y}^2}; \quad I2| = \sqrt{I|_{2x}^2 + I|_{2y}^2} \tag{6}$$

The electromagnetic moment is determined from the equation:

$$T_e = \frac{3}{2}pL_m\left(\overline{I_2} \cdot \overline{I_1}\right) = 1.5pL_m\left(I_{2x}I_{1y} - I_{2y}I_{1x}\right) \tag{7}$$

The energy properties of the MIS are determined after the calculation of the currents from the equations according to the following expressions.

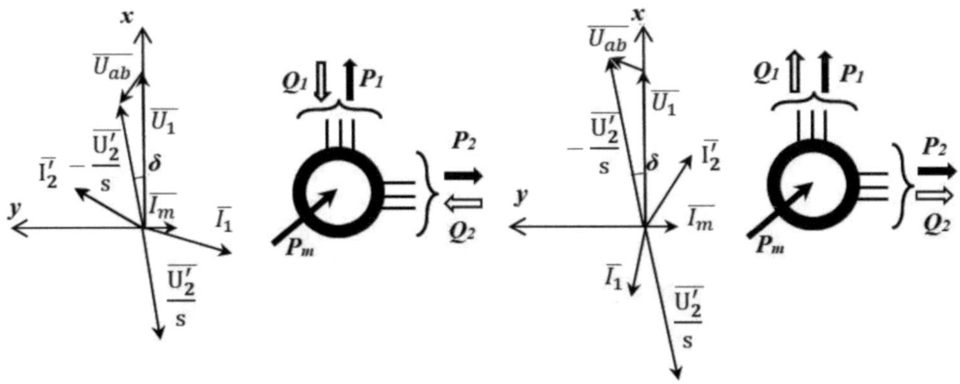

Figure 16.
Vector diagram and energy diagram of DFD in generator mode at n > n_o.

Slide	Generator mode
s > 0	$P_1 + P_m = -P_1 + \Delta P$
s < 0	$P_m = -P_1 + \Delta P - P_2$

Table 1.
Balance of DFIA active powers depending on operating modes.

$$P_1 = 1.5U_1 I_{1x};\ Q_1 = -1.5U_1 I_{1y};\ Q_2 = 1.5(U_{2y}I_{2x} - U_{2x}I_{2y});$$
$$P_m = T_e\omega_m = T_e\frac{(1-s)\omega_1}{p};\ P_2 = 1.5(U_{2x}I_{2x} + U_{2y}I_{2y} \tag{8}$$

Electromagnetic losses are calculated by the expression

$$\Delta P = I_1^2 R_s + I_2^2 | R_r \tag{9}$$

These equations can be used to model DFIA in Mathlab (**Figures 17–19**).
We present the main characteristics obtained on the model.
Assessing the properties of DFIA in the generator mode, we can draw the following conclusions:

1. The area of unstable operation is in a rather narrow zone of positive slides.

2. At $U_{2m} \leq U_1$ in the sliding range from 0 to −1, the reactive power consumed from the network is mainly positive (inductive).

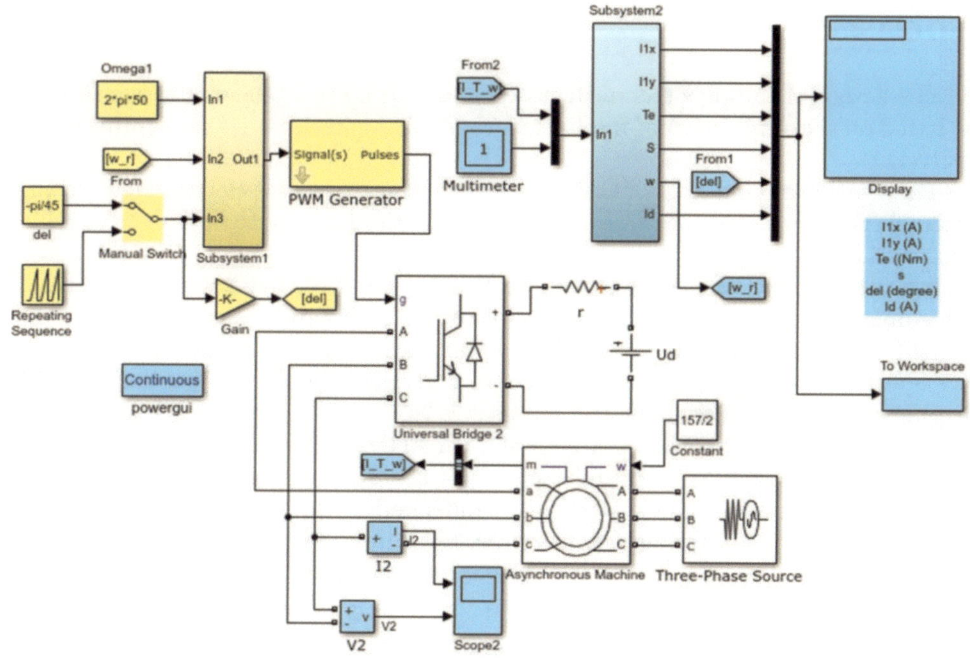

Figure 17.
Virtual laboratory setup for DFIA studies with independent controlling.

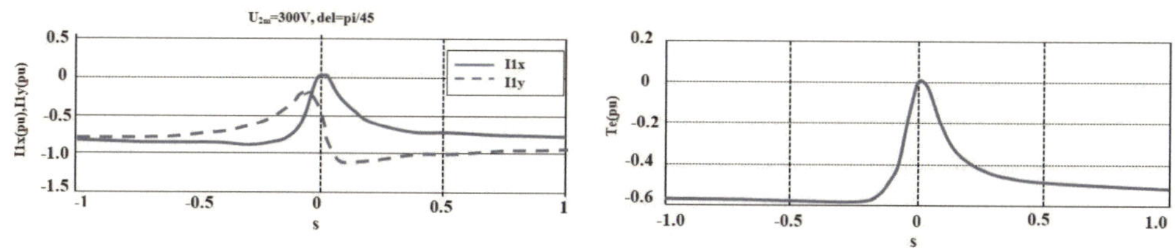

Figure 18.
Electromagnetic and electromechanical characteristics of DFIA with independent control in generator mode at $U_{2m} < U_1$.

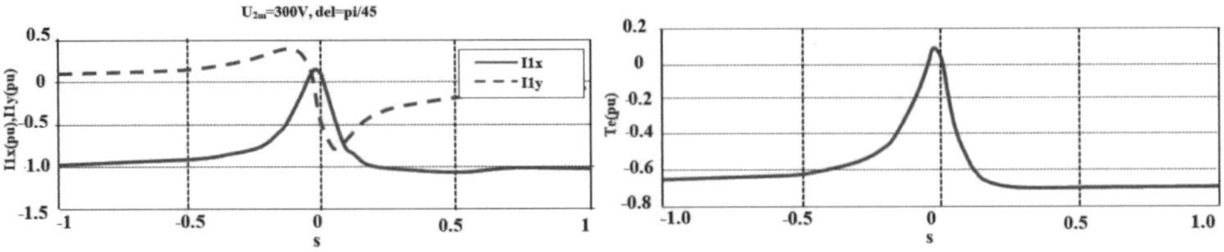

Figure 19.
Electromagnetic and electromechanical characteristics of DFIA with independent control in generator mode at $U_{2m} > U_1$.

3. At $U_2 > U_1$ in the sliding range from 0 to −1, the reactive power consumed from the network is always negative (capacitive).

4. In case of significant sliding, the DFIA is a source of torque.

This property is extremely useful when performing the DFIA generator function when the shaft speed varies widely.

6. DFIA design basics

The design of complex technical systems such as DFIA in the digital age should be based on the application of nonlinear programming techniques for synthesis and analysis.

There are a large number of CAD systems for the analysis of electromagnetic and thermal state of electric machines. Application of the Ansys software package will be shown below.

For the synthesis of electric machines, there are no fully developed techniques. Developers are forced to form a design system for each new type of machine.

DFIA operates in a wide range of mechanical and electrical loads. For its effective operation in all modes, it is necessary to determine the optimal geometry of the magnetic system and the anchor windings.

It should be noted that the problem of optimal design of electric machines remains quite complex and not fully solved [5, 12–18]. This is due to the choice of optimization method, the formalization of independent variables and the definition of quality indicators of the electric machine. Until now, the optimal design problems remain very important and relevant in electrical engineering.

6.1 Problem statement of optimal design of DFIA

Optimizations should be understood as the process of selecting the best option for a large number of possible options. Indicators of the quality of the best option are the optimality criteria. As a rule, there are several optimality criteria in the computational model, and in the most General case there is a need to solve a multi-criteria problem. It should be noted that the solution of multi-criteria problem is quite difficult. This is due to the fact that the optimality criteria are in a contradictory relationship. Improvement of one criterion leads to deterioration of others. So the increase in efficiency leads to an increase in the mass and volume of the product, the improvement of the output voltage parameters to the complication of electronics, cost reduction to a decrease in reliability. The experience of optimal design in electrical engineering shows that the task of multi-criteria optimization is quite complex. It is poorly formalized. The existing methods of its solution are

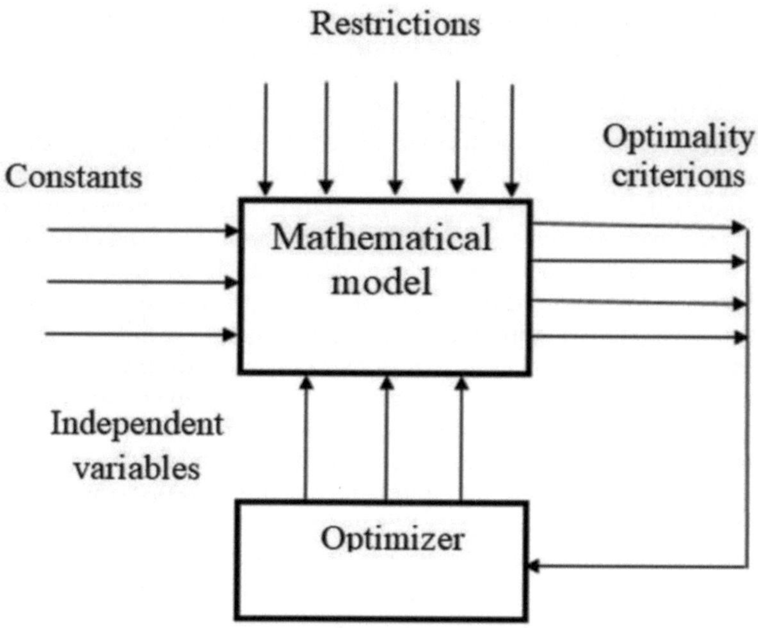

Figure 20.
Block diagram of the single-criterion optimization problem.

subjective and the success of their application depends on the experience of the developer. For this reason, we formulate a one-criterion optimization problem for the generator. Practice shows that such problems are solved successfully and efficiently. It should be noted that the optimality criterion can be changed depending on the current project situation.

We formulate the problem of single-criterion optimization of DFIA in the classical formulation [2, 3]: for the given parameters (specific performance, materials, initial data of the technical task), under the given restrictions, it is necessary, making a search of independent variables by a certain algorithm, to determine the geometry that would provide the extreme value of the selected criteria. There can be several optimality criteria, but depending on the project situation one is chosen for optimization.

The flowchart of the single-criterion optimization problem is shown in **Figure 20**.

The data of the technical task are taken as constants in the problem.

As constraints are the technical requirements for dimensions and permissible loads.

As criteria of optimality it is possible to accept the quality indicators which have received the greatest distribution in practice, for example: the minimum mass of active materials; the minimum volume of magnetic system, the maximum possible at the set electromagnetic loadings of efficiency.

The choice of independent variables is a separate and rather complex task.

6.2 Selecting independent variables for the optimization problem

The independent variables for the optimization problem must meet the following basic requirements:

- they should be visual and have a clear physical meaning;
- they should be effective, that is, their change should bring the calculation closer to the optimal design in the minimum number of iterations;

- they should have clear boundaries of change, it is desirable that the optimum is not on the border. This will make it difficult to choose optimization methods.

Analysis of the characteristics of the DFIA generator shows that its energy efficiency and specific energy performance are less dependent on specific linear dimensions, such as the height and width of the slot, the size of the back of the stator and inductor. To a greater extent, the energy of the machine depends on the ratio of the areas of the active zones, for example, the ratio of the cross-sectional area of the anchor to the cross-sectional area of the machine, the ratio of the area of the slotted zone to the area of the anchor, the ratio of the area of the slots to the area of the slotted zone, etc. Selecting these relative parameters as independent variables is very convenient. They have clear boundaries of change from 0 to 1, show the optimal ratio of the active zones involved in the energy conversion; their optimal values vary in a small range when the size changes over a wide range.

The idea of using these variables is not new. It was used for the synthesis of traction motors of rolling stock [22], and later, for the development of design methods of valve torque DC motors [3]. However, for dual power generators, this technique has not previously been used. Let us give these variables the term "generalized variables", which reflect their physical nature, and define them for DFIA.

6.3 Definition of generalized variables for DFIA

For the asynchronous-synchronous machine, it is convenient to allocate 6 generalized variables:

1. Variable f_s shows how much of the cross-sectional area of the electric machine without shaft is occupied by the total machine cross section (**Figure 21**);

$$f_s = \frac{S_{ring}}{S_{circle}} = \frac{S_{machine\ cross\ section\ without\ shaft}}{S_{total\ machine\ cross\ section}}; \tag{10}$$

2. Variable f_a reflects the ratio of area of the rotor cross section by the area of total machine cross section (**Figure 22**)

$$f_{\mathrm{a}} = \frac{S_{rotor\ cross\ section}}{S_{total\ machine\ cross\_section}}; \tag{11}$$

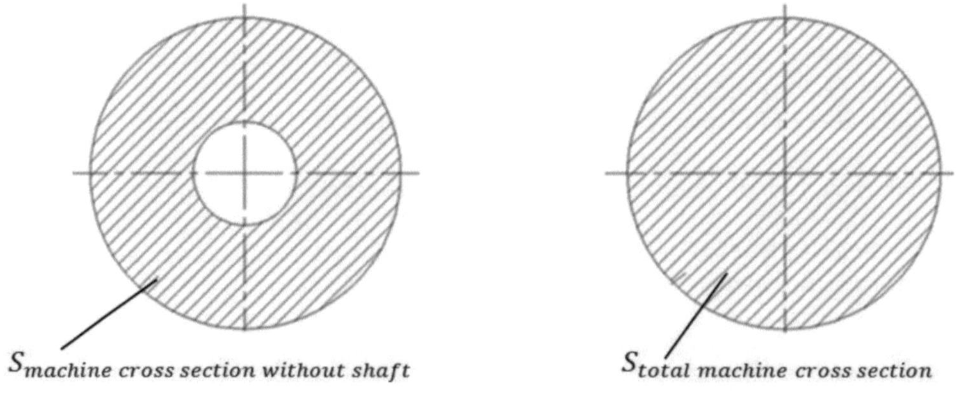

Figure 21.
Variable f_s.

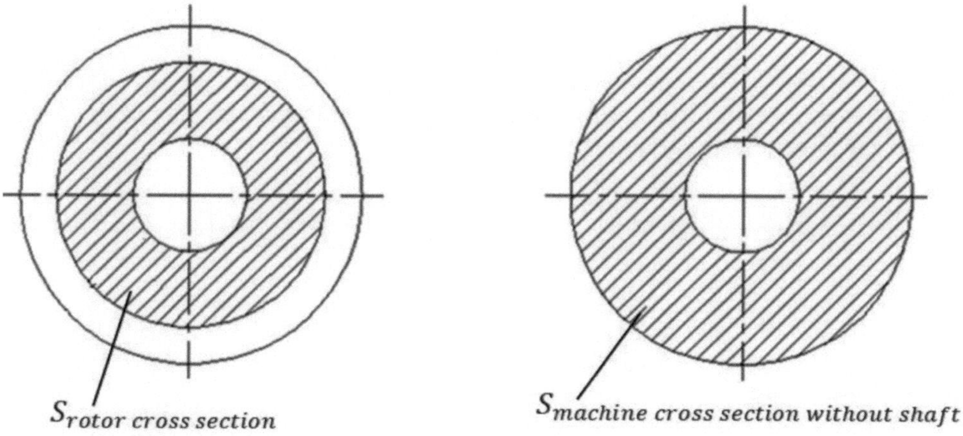

Figure 22.
Variable f_a.

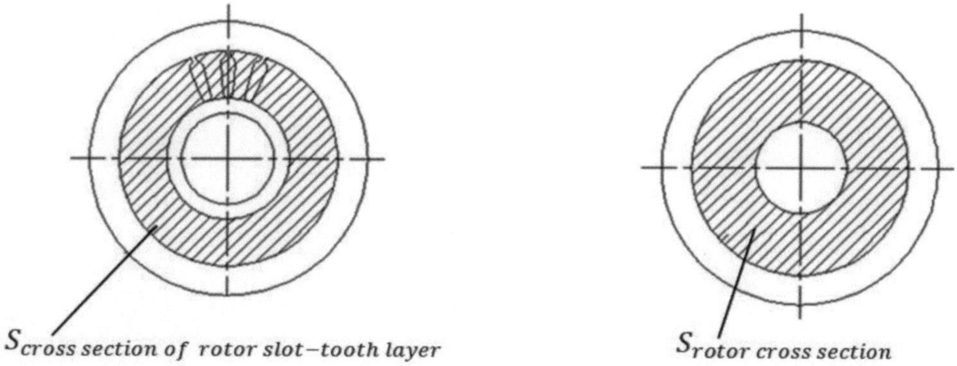

Figure 23.
Variable f_{es}.

3. Variable f_{er} shows the ratio of the area of cross section of rotor slot-tooth layer by the area of rotor cross section (**Figure 23**);

$$f_{es} = \frac{S_{cross\ section\ of\ \ rotor\ slot-tooth\ layer}}{S_{rotor\ cross\ section}}; \tag{12}$$

4. Variable f_{zr} reflects the ratio of the area of cross section of rotor slots by the area of cross section of rotor slot-tooth layer (**Figure 24**)

$$f_{zr} = \frac{S_{cross\ section\ of\ \ rotor\ slots}}{S_{cross\ section\ of\ rotor\ slot-tooth\ layer}}; \tag{13}$$

5. Variable f_{es} shows the ratio of the area of stator slot-tooth layer by the area of stator cross section (**Figure 25**);

$$f_{es} = \frac{S_{cross\ section\ of\ \ stator\ slot-tooth\ layer}}{S_{stator\ cross\ section}}; \tag{14}$$

6. Variable f_{zs} reflects the ratio of the area of cross section of stator slots by the area of cross section of stator slot-tooth layer (**Figure 26**);

$$f_{zs} = \frac{S_{cross\ section\ of\ \ stator\ slots}}{S_{cross\ section\ of\ \ stator\ slot-tooth\ layer}}; \tag{15}$$

Once again, we emphasize the visibility and clear physical meaning of the introduced generalized variables.

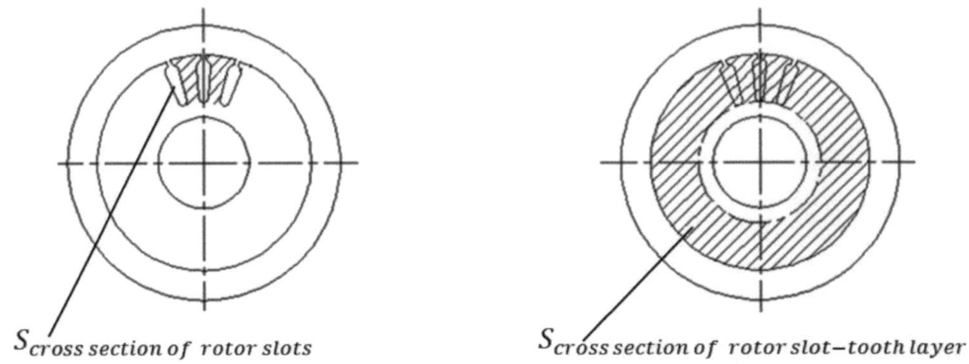

Figure 24.
Variable f_z.

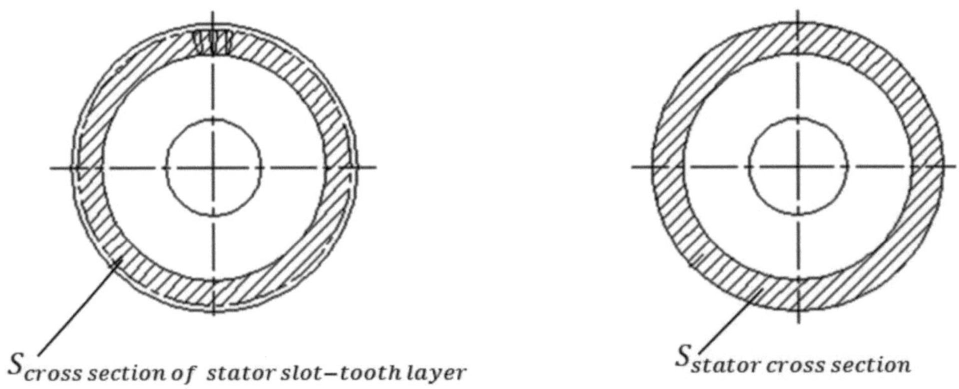

Figure 25.
Variable f_{er}.

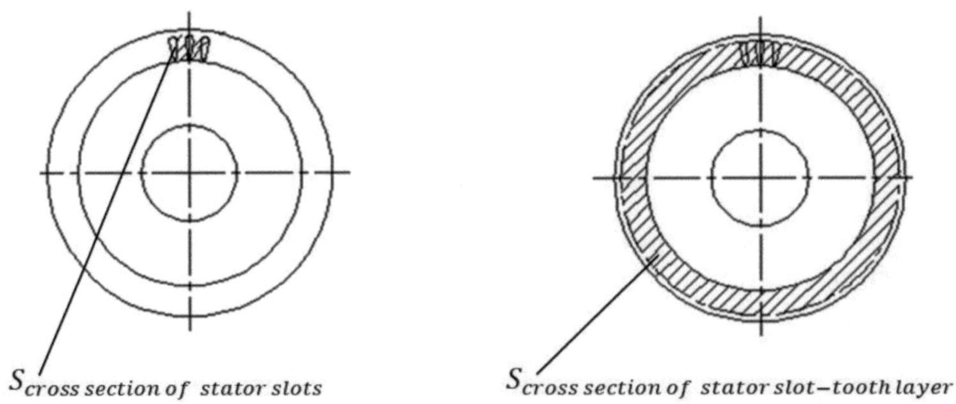

Figure 26.
Variable f_{zr}.

6.4 The determination of the geometry of DFIA using generalized variables

The above six generalized variables allow us to accurately calculate the basic dimensions of the magnetic circuit of the DFIA but for the product of this calculation it is necessary to set the initial parameter from which the further calculation will be made. As this parameter, it is most convenient to take the outer diameter of the stator of the electric machine, since most often this parameter is set by the terms of reference.

Taking as the initial parameter the outer diameter of the stator of the electric machine simple mathematical transformations allow from expressions (10),(11), (12), (13), (14), (15), print the equations that determine the basic geometric dimensions of the magnetic circuit of the double-feed machine (see **Figure 27**):

1. Variable f_s allows determining the diameter of the hole in the rotor (16):

$$Din = Dout \times \sqrt{1 - f_s}, \tag{16}$$

where *Dout* is the outer diameter of the stator;

2. Variable f_a allows determining the diameter of the rotor Da (17):

$$Da = Dout \times \sqrt{1 - f_s + f_a \times f_s}; \tag{17}$$

3. The variable f_{er} allows determining the height of the groove of the rotor h_{zr} (18):

$$h_{zr} = 0.5 \times Dout \times \left(\sqrt{1 - f_s + f_a \times f_s} - \sqrt{1 - f_s + f_a \times f_{er} + f_a \times f_s};\right. \tag{18}$$

4. The variable f_{zr} allows defining the width of the groove of the rotorb_{zr} (19):

$$b_{zr} = \frac{\pi \times Dout \times f_{zr}}{2 \times z_p} * \left(\sqrt{1 - f_s + f_a \times f_s} + \sqrt{1 - f_s + f_a \times f_{er} + f_a \times f_s},\right. \tag{19}$$

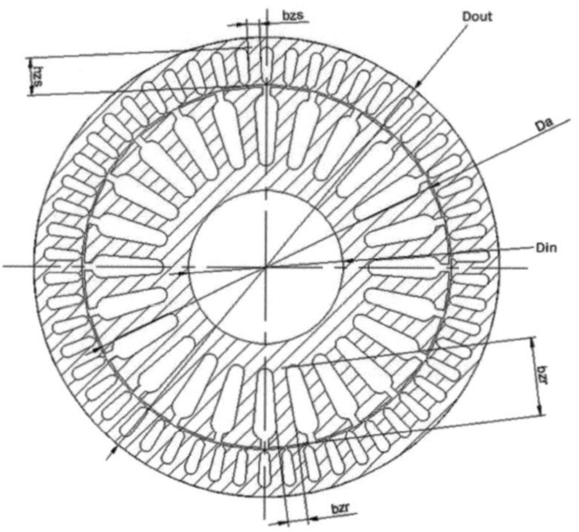

Figure 27.
Cross-sectional sketch of DFIA.

where z_p is the number of rotor slots;

5. The variable f_{es} allows determining the height of the stator slot h_{zs} (20):

$$h_{zs} = 0.5 \times \left(\sqrt{f_{es} \times Dout^2 - (Da + 2 \times \delta)^2 \times (f_{es} + 1)} - Da - 2 \times \delta, \qquad (20)$$

where δ is the air gap;

6. The variable f_{zs} allows determining the width of the stator groove b_{zs} (21):

$$b_{zs} = \frac{f_{zs} \times f_{es} \times \pi \times \left(Dout^2 - (Da + 2 \times \delta)^2\right)}{4 \times z_s \times h_{zs}}, \qquad (21)$$

where z_s is number of stator slots.

Thus, for a known outer diameter and given generalized variables defined by expressions (10–15), the transverse geometry of DFIA can be uniquely determined. On the basis of the accepted variables, a mathematical model of the generator was developed, which was used to optimize the geometry of a number of wind power plants based on DFIA. As a block optimizer used an algorithm that combines the method of coordinate-wise descent Gauss-Seidel with the Fibonacci method.

The main advantage of the generalized variables is their visibility, due to the fact that they have a clear physical meaning and can independently act as criteria for the quality of the electric machine. The second advantage is that the generalized variables have a fixed range of changes [0;1], which allows to apply optimization algorithms to the mathematical model [5, 18], built on the basis of these variables.

On the basis of this technique and with the use of the algorithm, DFIA with output power of 10 kV, line voltage 380 V and synchronous speed of 50 rpm with the required optimality criteria—maximum efficiency and minimum weight—was calculated . The calculation time of the program written in Delphi for six independent generalized variables was about 30 seconds, which is quite acceptable.

7. Analysis of the results of geometry optimization

At the stage of DFIA synthesis the method of substitution schemes was used, which contains equations in integral form. This is due to the fact that a simplified calculation model is required for optimization. This model should allow the calculation of a large number of options to select the best option.

A thorough analysis of the results after optimization is required. This analysis should use accurate methods. The finite element method is such an accurate method [2–4, 8–18, 23].

The software package Ansys Electronics Desktop allows you to perform analysis of electromagnetic and fields in an electric machine. The application of the program for the analysis of DFIA is shown below.

The data of alternator is:

Rated output power, (kW)	10
Rated voltage, (V)	380
Frequency, (Hz)	50
Rated speed, (RPM)	60

Mass, (kg)	1200
Outline sizes (mm*mm)	1260*320

The calculation in Maxwell Ansys can be divided into three stages:

1. Analysis of the model using the RMxprt;

2. Analysis of the model using Maxwell2D on the basis of already created models in RMxprt;

3. Analysis of the model using the Maxwell3D on the basis of the model created in RMxprt.

The program Ansys Electronics Desktop has a very developed interface. This makes it easier to use.

At the first stage, when working with RMxprt, it is enough to fill in special forms. They include geometrical dimensions, winding data and nominal parameters of the machine. The mathematical model of calculation is made on the basis of a method of schemes of substitution therefore are carried out within 1 min. The program calculates all the necessary parameters and characteristics.

At the second stage, the program automatically creates a 2D model of the generator, splits it into finite elements, sets the boundary conditions and current loads. To simplify the calculations, the program determines the axial symmetry and solves the problem of calculating the field taking into account the periodicity condition. The resulting characteristics are obtained by multiplying the results by the number of symmetric fragments.

To test the generator, the following numerical experiment was carried out: the different rotation frequency of the generator shaft (0, 10, 60 and 100) rpm was set.

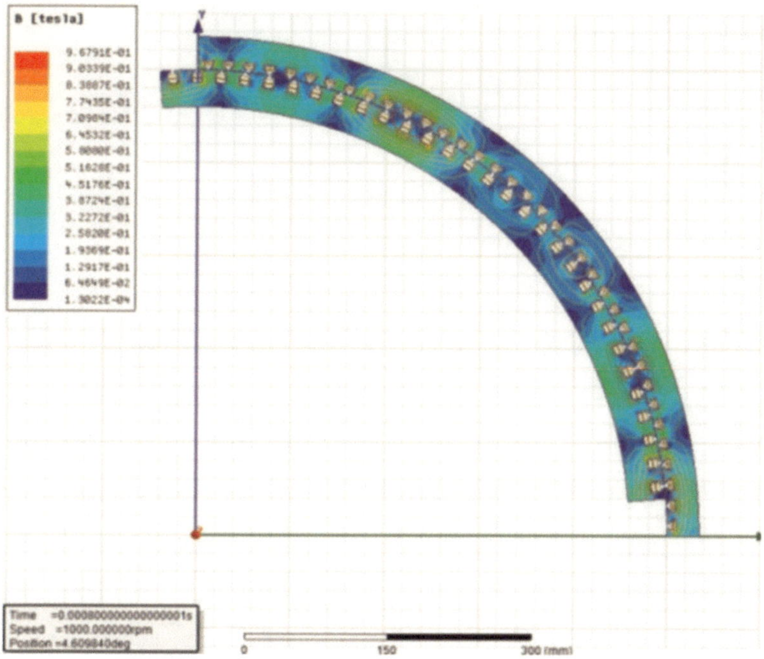

Figure 28.
The analysis of the magnetic field at frequency of rotation of 10 rpm.

For these speeds, the frequency of the excitation winding supply and the current in the excitation winding were selected so that the nominal voltage was obtained at the generator output.

The results of the analysis are presented below (**Figures 28–31**).

Tables 2 and **3** show the main electrical parameters of the generator obtained experimentally.

The results of the numerical experiment show that the output voltage and power have stable values when the rotation frequency of the generator shaft changes from a braked state to 10 RPM. It is very important for wind turbines.

Thus, the main problem of wind power to obtain standard parameters of electricity at different wind speeds is solved through the use of DFIA.

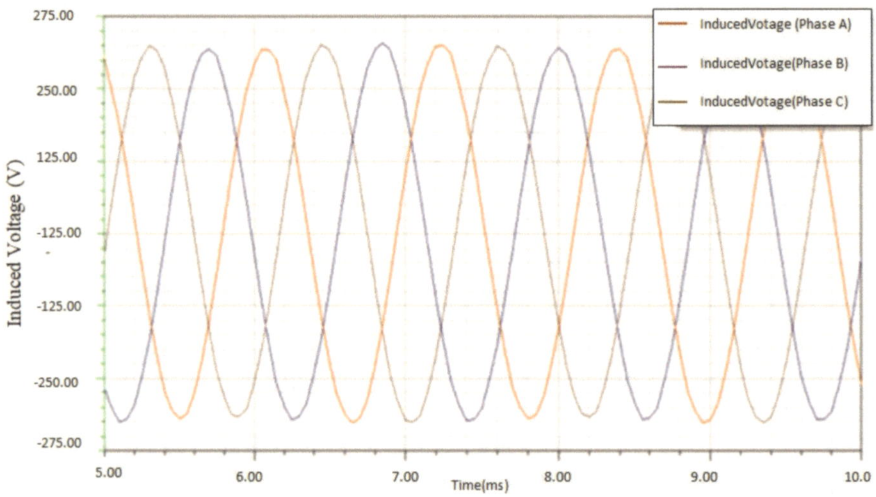

Figure 29.
EMF of anchor winding (10 rpm).

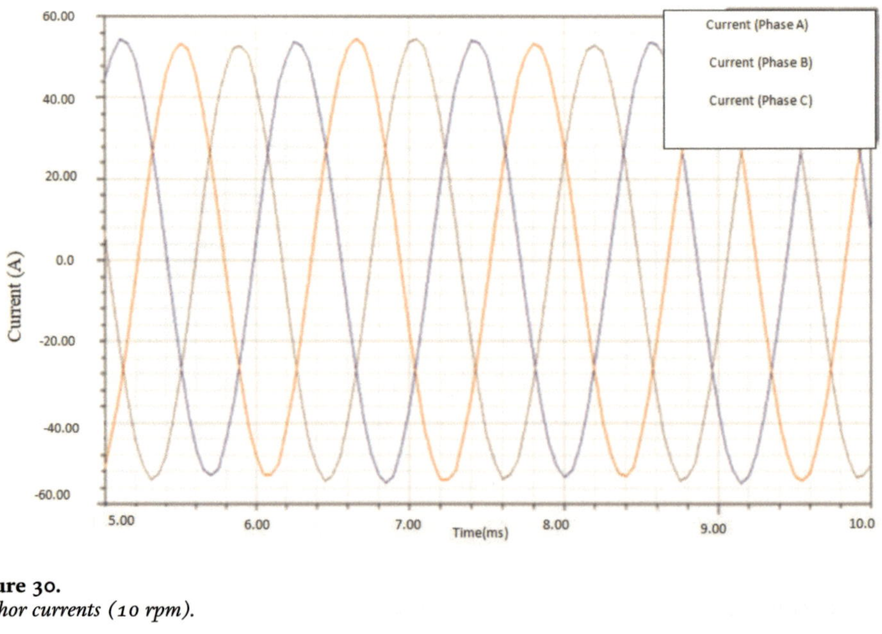

Figure 30.
Anchor currents (10 rpm).

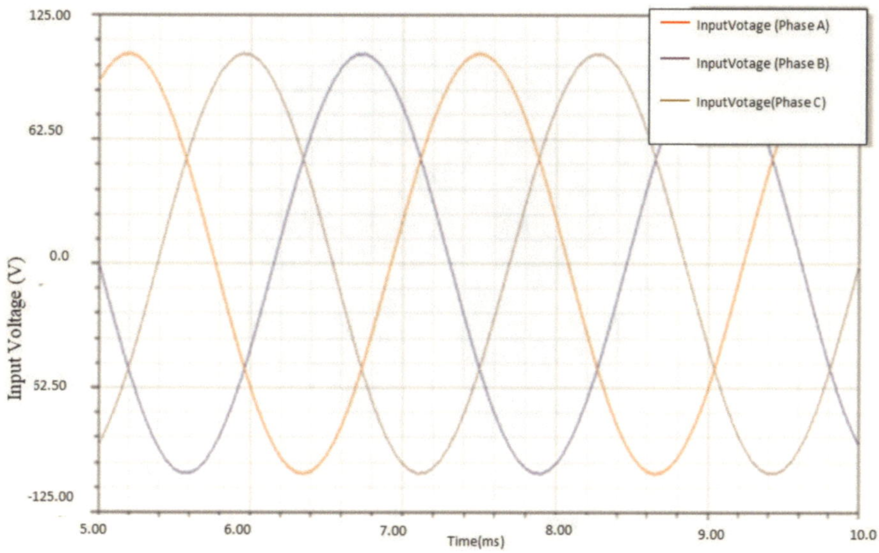

Figure 31.
The voltage of the exitating winding (10 rpm).

Rotational speed, rpm	Stator voltage (RMS), V	Load current (RMS), A	Excitation winding voltage (RMS), V	Excitation winding current(RMS), A
0	236	15.05	148.0	26.3
10	235	14.95	137.0	26.25
60	232	14.43	74.5	26.0
100	233	14.59	2.8	31.85

Table 2.
Generator voltages and currents depending on speed.

Rotational speed, rpm	Output power, kW	Excitation power, kW	Mechanical power on shaft, kW	Total power, kW	Efficiency
0	10.65	11.67	0.00	11.54	0.912
10	10.54	10.78	0.55	21.03	0.93
60	10.05	5.81	4.55	19.72	0.97
100	10.20	0.267	10.03	19.41	0.99

Table 3.
Generator p depending on speed.

8. The solving of the problem of an accumulation with using the concept of DFIA

Wind power plants must be highly reliable. This is due to the fact that the generator is located at high altitude, works in different climatic zones, and it is very difficult to carry out repair work in these conditions.

This requirement makes the supply of electricity to the rotor without sliding contact.

In the proposed generator, a second built-in electric machine is used to solve this problem.

Figure 32.
DFIA design with built-in rechargeable batteries.

But it is possible to apply another technical solution. Batteries can be used to power the electrical circuits of the rotor if placed on a rotating part (**Figure 32**). In this case, the exiting alternator can be excluded. The battery pack will additionally function as a storage device [4].

In practice it is impossible to use windmills without energy storage devices. It is necessary for us to have a stable energy supply but wind has unstable action.

Nowadays the storage batteries are the acceptable energy storage devices taking in consideration the reliability and cost. The double-fed alternator allows combining two concepts (creating the standard energy and storage energy) at one device. In sketch, it is possible to see that the rotor has enough place for allocation the batteries (**Figure 32**).

These batteries can solve two problems. They are:

- The electric energy storage.

- The supplying the electric energy to electronic control system. This solution allows extending the limits of the windmill working operation. Indeed, for short time, it is possible to use windmill with fixed rotor when we have no wind.

9. Conclusion

Nowadays the development of windmill industry is holding back by one serious problem. It is very difficult to transform with high efficiency the nonstandard energy to the standard energy. The using of the induction alternators or the synchronous alternators cannot solve this problem with good results. As a rule in this case we have low efficiency, complex design and high total cost. The problem is becoming more difficult when you make attempt to build the row of windmills from small power to big power (10 kW–1 MW). The concept of the DFIA proves to be very effective for solving windmills problems. It is possible to transform the nonstandard energy in to the standard energy with high efficiency. Windmills may operate as stand-alone or grid tied. It is possible to build the row of windmills with the same design. But the main reason of the application the DFIA is the low total cost of windmill and electronic control system. The new design of this concept is presented. The concept is confirmed by the complex engineering calculations and the manufacturing of the prototypes.

This concept can be used for design of windmill row with output power up to 1 MW.

Author details

Sergey Gandzha* and Anton Kotov
South Ural State University, Chelyabinsk, Russia

*Address all correspondence to: gandja_sa@mail.ru

References

[1] Altshuller GS. Find an Idea. 3-e Izd., EXT. Petrozavodsk: Scandinavia; 2003. (in Russian)

[2] Gandzha S, Aminov D, Kosimov B. Application of digital twins Technology for Analysis of brushless electric machines with axial magnetic flux. Proceedings-2018. Global Smart Industry Conference CloSIC. 2018

[3] Gandzha S, Aminov D, Kosimov B. Design of Brushless Electric Machine with axial magnetic flux based on the use of Nomograms. In: Proceedings - 2018 International Ural Conference on Green Energy. UralCon; 2018

[4] Gandzha S, Belonozhko A. Development of electrical energy storage device using direct-acting fuel cells based on methanol. In: Proceedings - 2018 International Ural Conference on Green Energy. UralCon; 2018

[5] Stemmler H, Omlin A. Converter controlled fixed-frequency variable-speed motor/alternator. Presented at the IPEC '95, Japan; 1995

[6] Bogalecka E. Dynamics of the power control of double fed induction alternator connected to the soft grid. In: Proc. IEEE Int. Symp. Ind. Electron. Budapest; 1993. pp. 509-513

[7] Lauw HK, Weigand CH, Marckx DA. Variable-Speed Wind System Design: Final Report. Washington, DC, Rep. DOE/BP/99893-TI: U.S. Dept. of Energy; 1993

[8] Gandzha S, Kiessh I, Aminov D. Development of engineering technique for calculating magnet systems with permanent magnets. In: Proceedings - 2018 International Conference on Industrial Engineering, Applications and Manufacturing. ICIEAM; 2018

[9] Gandzha S, Kiessh I. Selection of winding commutation for axial gap machines with any phases. In: Proceedings - 2018 International Conference on Industrial Engineering, Applications and Manufacturing. ICIEAM; 2018

[10] Gandzha S, Andrey S, Andrey M, Kiessh I. The design of the low-speed brushless motor for the winch which operates in see-water. In: International Multidisciplinary Scientific GeoConference Surveying Geology and Mining Ecology Management. SGEM; 2017

[11] Gandzha S, Kiessh I. The high-speed axial gap electric alternator is the best solution for a gas turbine engine. In: International Multidisciplinary Scientific GeoConference Surveying Geology and Mining Ecology Management. SGEM; 2017

[12] Gandzha S. The application of the double-fed alternator for the solving of the wind power problems. In: International Multidisciplinary Scientific GeoConference Surveying Geology and Mining Ecology Management. SGEM; 2017

[13] Gandzha SA, Sogrin AI, Kiessh IE. The comparative analysis of permanent magnet electric machines with integer and fractional number of slots per pole and phase. Procedia Engineering. 2015; **129**:408-414

[14] Gandzha SA, Kiessh IE. Application brushless machines with combine excitation for a small and medium power windmills. Procedia Engineering. 2015;**129**:191-194

[15] Gandzha S, Kiesh I. A proposal of doubly-fed alternator for windmill application. In: 2016 2nd International Conference on Industrial Engineering. Applications and Manufacturing (ICIEAM). IEEE Conferences; 2016. pp. 1-3

[16] Gandzha SA, Kiessh IE. Varible speed power. Procedia Engineering. 2015;**129**:731-735

[17] Gandzha SA, Sogrin AI, Kiessh IE. The comparative analysis of permanent magnet electric machines with integer and fractional number of slots per pole and phase. Procedia Engineering. 2015; **129**:408-414

[18] Gandzha SA. Modelling of permanent magnet direct current motor with electromagnetic reduction. In: Gandzha SA, editor. Collection of Papers of Software Users Sixth Conference CAD_FEMGmbH (20–21 April 2006). Moscow; 2006. pp. 358-360

[19] Kurochka AL. Synthesis of optimal machines of direct and pulsating current on the basis of aggregated variables. In: Kurochka AL, editor. Electromechanics. Vol. 6. Izv. Higher. Educational Institution; 1976. pp. 608-617. (in Russian)

[20] Botvinnik MM. Asynchronous Synchronous Machine. Moscow, Leningrad: Gosenergoizdat; 1960. 70 p. (in Russian)

[21] Slezhanovsky OV. In: Slezhanovsky OV, Dackowski LH, Kuznetsov IS, et al., editors. Systems of Subordinate Regulation AC Motor Drive with Brushless DC Converters. Moscow: Energoatomizdat; 1983. (in Russian)

[22] Herman-Galkin SG. Virtual Laboratories of Semiconductor Systems in Mathlab-Simulink Environment: Textbook. LAN Publishing House; 2013. 448 p. (in Russian)

[23] Cingoski V, Murakawa R, Kaneda K, Yamashita H. Automatic mesh generation in finite element analysis using dynamic bubble system. Jornal of Applied Physics. 1997;**81**(8 Pt 2): 4085-4087

Chapter 2

Aerodynamics of Low-Rise Buildings: Challenges and Recent Advances in Experimental and Computational Methods

Aly Mousaad Aly, Faiaz Khaled and Hamzeh Gol-Zaroudi

Abstract

Buildings are bluff bodies, compared to streamline objects, such as airfoil. Wind flow over buildings leads to separation and hence a complex spatial and temporal mechanism that governs the nature and intensity of aerodynamic forces. This complexity mainly comes from the transient nature of incident turbulent winds and the fluctuating flow pattern in the separation bubble. The study of building aerodynamics is vital for the evaluation of cladding pressures, drag, shear, and uplift forces that are essential for safe and economic design. Flow separation makes it challenging to estimate loads without referring to direct physical and/or computational simulation. For several decades, aerodynamic testing has been employed for the estimation of wind pressures and forces on buildings. However, for residential homes and low-rise buildings, it has been always a challenge to predict full-scale pressures by traditional wind tunnel testing, as per the lack of large turbulence and Reynolds number effects, among other factors. The mismatch in flow physics makes it difficult to scale up wind-induced loads as the process can be highly nonlinear, which is the case when full-scale pressure coefficients do not meet those from small-scale aerodynamic testing. This chapter presents the challenges in the modeling and evaluation of aerodynamic forces on low-rise buildings, along with recent advances in both experimental and computational methods.

Keywords: aerodynamics, wind engineering, open-jet testing, wind tunnel, atmospheric boundary layer, low-rise buildings

1. Introduction

Researchers and engineering practitioners are attentive to understanding the behavior of structures under the effects of various loading patterns and conditions, to enhance their lifetime performance. Wind forces can threaten the safety of structures if their effects are underestimated; therefore, it is crucial to properly simulate and assess wind effects on civil engineering structures in order to achieve optimal and resilient designs that can maintain accessibility and functionality after natural disasters. Due to climate change and its consequences, the patterns of extreme winds and hurricane occurrence have been altered [1–3]. As a result, wind loads are becoming important in the analysis and design of buildings, especially in hurricane active regions. To put it into perspective, in most parts of the United States, especially in the east coast and the southern region, hurricanes and severe windstorms hit and bring widespread damage to buildings and other types of structures. The associated losses are estimated in billion dollars. The normalized hurricane-induced damage in the United States, between 1900 and 2005 (106 years of record), was estimated at about $10 billion (normalized to 2005 USD) [4]. Damage records totaling $265 billion were set by hurricanes Maria, Harvey, and Irma [5].

Due to the population growth, coastal zones are being more and more concentrated with residential buildings. These buildings are mostly light and low-rise, constructed from wooden materials, with different aerodynamic performance compared to high-rise buildings and residential homes. The American Society of Civil Engineers (ASCE) design standard defines a low-rise building to have an average roof height that is less than its lateral dimension; however the building should not exceed 18.3 m [6]. The majority of failures in low-rise buildings are reported because of strong wind effects on their envelope and specifically on roof panels [7]. **Figure 1(a)** shows a total failure of a low-rise building induced by hurricane Sandy in New York in 2012 [8]. The building envelope experienced significant loads from hurricane winds and lost its load path connections. In other scenarios, once part of a roof is breached during high winds, it facilitates the penetration of rainwater which can be harmful to interior properties and may cause serious problems to the building and loss of contents. **Figure 1(b)** shows severe roof damage during Hurricane Katrina in Lake Charles, New Orleans, in 2005 [9].

Examination of post-disaster surveys indicates initiation of damage through failure of roof components under extreme wind events. Earlier studies confirm the presence of extreme negative pressures at corners, ridges, and leading edges of roofs. The performance of roofs in low-rise buildings can differ significantly during a windstorm according to the shape of roof and its dimension. For instance, large roofs in industrial buildings may behave differently, compared to those of small roofs in a single-family low-rise building which can lead to different damage patterns to the building envelope [10–12]. In large roofs, the correlations among pressures acting at different roof locations are usually low [13]. In large roofs of light metal industrial buildings, leading edge failure usually occurs due to poor attachment of metal sheathing in areas that are exposed to uplift wind forces. This weakness eventuates to progressive peeling of the roof membrane causing further damage to the whole integrity of the building envelope.

The components and claddings in small roofs are usually exposed to damage during windstorms, due to local fluctuating negative pressures (uplift effects) due

(a) (b)

Figure 1.
Hurricane-induced damage: (a) complete collapse of a residential home induced by hurricane Sandy, New York, 2012 [8] and (b) severe roof damage by hurricane Rita in Lake Charles, in 2005 [9].

to flow separation, especially at roof edges and corners. **Figure 2** represents wind flow around a residential building [13]. The flow separates at sharp edges and re-attaches again in a fluctuating manner within the separation zones at a distance that is called separation bubble length, leading to uplift forces on the roof surface. The stagnation point is also specified in the windward wall, where the along-wind velocity is zero. **Figure 3** shows homes damaged by Hurricane Andrew in 1992 as a result of low pressures on the roof; and as a result, the shingles and sheathings were blown off due to high uplift forces. Referring to **Figure 2**, now it is shown that the separation bubble effects and the flow detachment are the main causes of these damage patterns of roof coverings which are a representation of roof areas under uplift forces. To fully understand windstorm effects on low-rise and residential buildings, it is essential to replicate the physics by experimental and computational methods. There are two important requirements: (1) correct reproduction of the main characteristic in the atmospheric boundary layer (ABL) and (2) aerodynamic testing at proper scales.

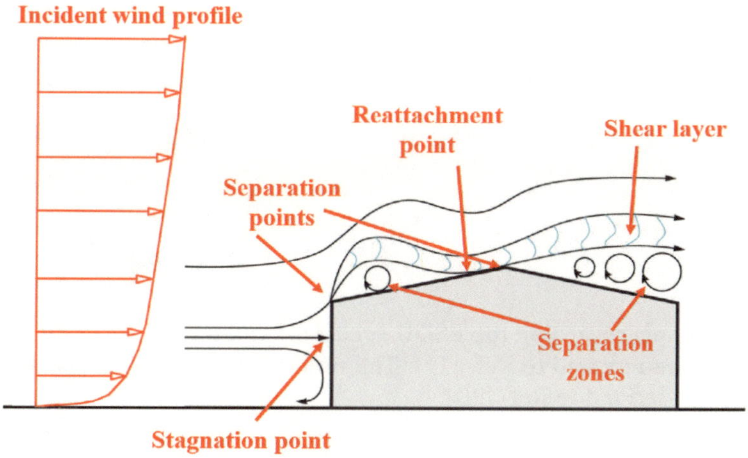

Figure 2.
Fluctuating flow separation and re-attachment (adapted from Ref. [14]).

Figure 3.
Homes damaged by hurricane Andrew in 1992 [15].

2. Atmospheric boundary layer characteristics

The variation of the mean velocity profile with height can be different over different terrain conditions depending on the friction effects from the earth's surface and the value of roughness length. **Figure 4** shows a schematic of different mean wind profiles over various topographical conditions of a dense urban area, suburban terrain, and over sea surfaces. In **Figure 4**, higher velocity is anticipated in lower altitudes on sea surfaces than the gradient wind in a dense city center.

After recording time series of wind velocity in the lab or in the field, the turbulence spectrum can be obtained accordingly. For the validation of the turbulence spectrum, theoretical spectra are usually used. The Kaimal spectrum is one of the widely used spectra, which is defined as follows [17]:

$$\frac{f S_{aa}(f)}{u_*^2} = \frac{An}{(1+Bn)^{5/3}} \tag{1}$$

in which f is nU/z. One can obtain the spectrum, S_{uu}, in the along-wind direction by considering A to be 105 and B to be 33 [14, 18]. For the lateral and vertical spectra, different values for the parameters A and B are suggested [14, 18].

The Engineering Science Data Unit (ESDU) spectrum is proposed based on a new von Karman spectrum, covering the full frequency range, as follows [19]:

$$\frac{f S_{uu}(f)}{\sigma_u^2} = \beta_1 \frac{2.987 n_u/\alpha}{\left[1+(2\pi n_u/\alpha)^2\right]^{5/6}} + \beta_2 \frac{1.294 n_u/\alpha}{\left[1+(\pi n_u/\alpha)^2\right]^{5/6}} F_1 \tag{2}$$

For more details regarding the ESDU spectrum and definition of different terms, readers are referred to Ref. [19]. The nondimensional cross-spectrum of u-component is defined in Ref. [20]:

$$\tilde{S}_u^c(\zeta, n) = \frac{S_u^c(\zeta, n)}{\sqrt{S_{uA}(n)}\sqrt{S_{uB}(n)}}, \tag{3}$$

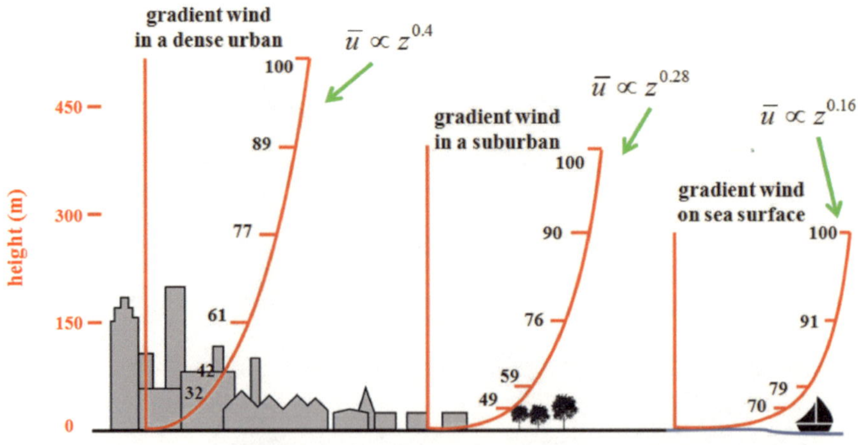

Figure 4.
Mean wind speed profiles over different terrains according to Davenport's power law profiles (adapted from Ref. [16]).

$$S_u^c(\zeta, n) = \int_{-\infty}^{\infty} R_u^c(\zeta, \tau) \exp(-j2\pi n\tau) d\tau. \tag{4}$$

where $R_u^c(\zeta, \tau) = E[u_A(t)u_B(t+\tau)]$, $S_{uA}(n)$, and $S_{uB}(n)$ are power spectra at two points, A and B, respectively; n is the frequency; and ζ is the distance between the two points A and B. The cross-spectrum of Davenport is defined in Ref. [21]:

Davenport:

$$\tilde{S}_u^c(\zeta, n) = \exp\left(-k_r \frac{|n\zeta|}{\overline{U}}\right), \tag{5}$$

Maeda and Makino:

$$\tilde{S}_u^c(\zeta, n) = \exp(-k_1\theta).(1 - k_2\theta^2), \tag{6}$$

where $k_r = 13(\zeta/z_m)^{0.4}, z_m = 0.5(z_A + z_B), \theta = \left\{(0.747\zeta/{}^xL_u)^2 + (2\pi n\zeta/\overline{U})^2\right\}^{1/2}$, $k_1 = 1.0$, and $k_2 = 0.2$ [20].

The integral length scale of turbulence, L_u^x, is a measure of the size of the largest eddy in a turbulent flow [18]. Having the time history of along-wind velocity component at any height, L_u^x can be calculated using the approach described in Ref. [22]:

$$L_u^x = \left[\frac{E(f)U_{mean}}{4\overline{u}^2}\right]_{f\to 0} \tag{7}$$

where $\bar{u}$ is the standard deviation of the along-wind velocity component and $E(f)$ is the power spectral density. Studies show that the integral length scale of turbulence may decrease in the flow direction, due to the fact that larger eddies will usually dissipate energy into smaller eddies [23]. According to actual measurements, as the terrain roughness decreases, L_u^x increases with the height above ground [18]. To quantify these changes, the integral length scale formulation suggested by ESDU is defined as follows [19]:

$$L_u^{ESDU} = 25z^{0.35}z_0^{-0.063} \tag{8}$$

And Counihan formulation used by Refs. [24, 25]:

$$L_u{}^C = 300(z/300)^{0.46+0.074\ln z_0} \tag{9}$$

3. Aerodynamics of low-rise buildings

Bluff body aerodynamics, and in particular fluctuating pressures on low-rise buildings immersed in turbulent flows, are associated with the complex spatial and temporal nature of winds [26]. This complexity mainly comes from the transient nature of incident turbulent winds, and the fluctuating flow pattern in the separation bubble. The flow in the separated shear layer is associated with fluctuations in the velocity field leading to the evolution of instabilities. The flow physics are dependent on upstream turbulence intensity, integral length scale, as well as Reynolds number. The later makes it difficult to scale up loads based on pressure and force coefficients as the process can be highly nonlinear, which is the case, for example, when full-scale pressure coefficients do not meet those from small-scale aerodynamic testing (**Figure 5**). Not only free stream turbulence impacts the flow pattern around bluff bodies, but also it can impact the thickness and length of the wake, hence significantly altering aerodynamic pressures.

In order to propose mitigation alternatives to minimize damages induced by windstorms to low-rise buildings, it is vital to understand how peak loads and spatial correlation of pressures are developed. As a first step to understand this mechanism, a true simulation of flow characteristics in accordance to real full-scale winds is necessary. There are common and valuable resources for the physical investigation of wind effects on structures, including small-scale wind tunnel testing, large-scale testing an open-jet laboratory, and full-scale field measurements.

According to Ref. [27], at relatively large-scale wind tunnel models, it is very difficult to model the full turbulence spectrum, and only the high-frequency end is matched [28]. For instance, as described in Ref. [29], more than 50% discrepancies in wind tunnel aerodynamic measurements are realized from six reputable centers for roof corner pressure coefficients and peak wind-induced bending moment in structural frames of low-rise building models. Therefore, selecting an appropriate testing protocol, including model scale ratio, for physical testing to minimize discrepancies in aerodynamic loads is essential. This can be achieved by considering constraints on laboratory testing that limits producing the large-scale turbulence and the inherent issues with limited integral length scale [30].

The literature raises questions regarding the adequacy of predicting full-scale pressures on low-rise buildings tested in flows that lack the large-scale turbulence. For instance, although a good agreement was observed between a wind tunnel testing on a generic low-rise building and full-scale data, discrepancies were shown in reproducing the largest of peak pressure near roof edges [31]. **Figure 5** shows minimum pressure coefficients at a building corner and eave level for the full-scale

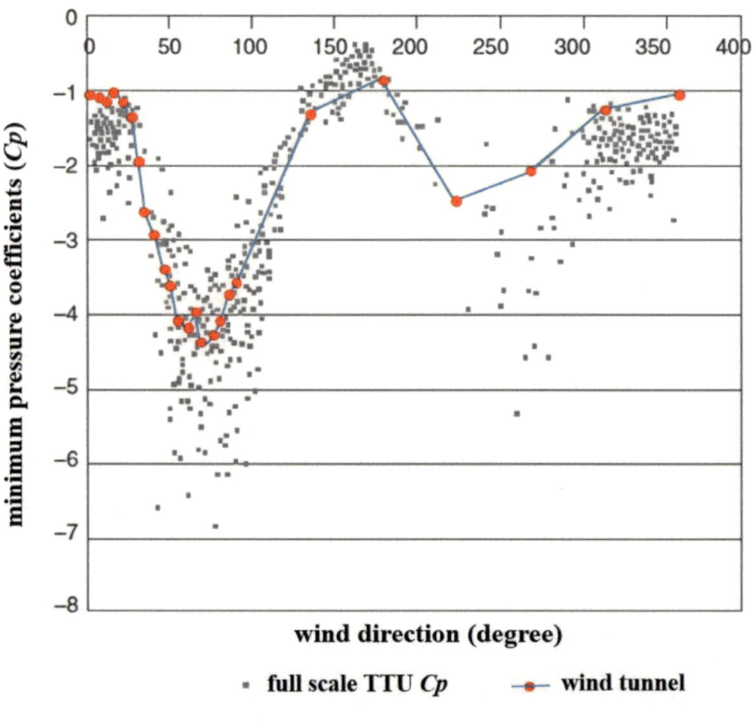

Figure 5.
Minimum pressures at building corner (adapted from Ref. [14]).

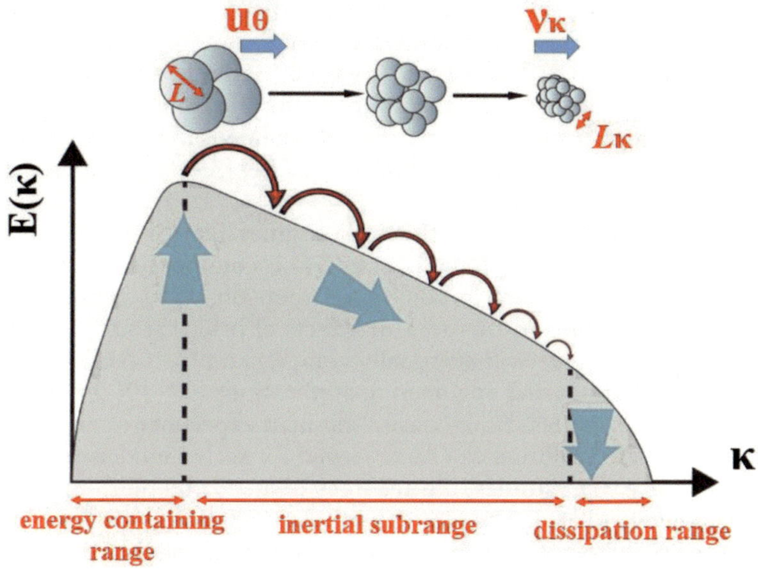

Figure 6.
Energy cascade in a turbulent flow (adapted from Ref. [32]).

Texas Tech University (TTU) experimental building, along with wind tunnel measurements [14]. The local peak pressures are weaker in wind tunnel testing than those at full-scale. For instance, at 65° wind direction angle, the wind tunnel reproduced minimum pressure coefficient of −4.3, while the full-scale field measurement is −6.8, and at 250° wind direction angle, wind tunnel shows −2.2, while the full-scale data shows −5.3. Therefore, there would be a major doubt on estimating the correct wind loads for design purposes based on wind tunnel testing. To describe this mismatch, first we need to define the concept of the energy cascade in a flow.

As depicted in **Figure 6**, the structure of a turbulent wind flow is constituted from a combination of large eddies and small eddies. In physical space, the large eddies are broken into smaller and smaller eddies with different spectral energy contents in various frequency ranges. In conventional wind tunnel testing, it can be challenging to appropriately reproduce low-frequency turbulence, which overwhelmingly contributes to the integral length scale and intensity of fluctuations. This leads to significant disparity among the wind tunnel flows and the target full-scale field flow conditions. As observed in **Figure 5**, this mismatch affected the local vorticity at edges and corners of a low-rise building model tested in a wind tunnel and resulted in local pressures weaker than those at full-scale. To alleviate these issues and to replicate the ABL flow characteristics for aerodynamics of buildings, advanced research in computational and experimental methods is essential.

4. Computational fluid dynamics in aerodynamics of buildings

In recent years, computational fluid dynamics (CFD) simulations have witnessed a spread use and applications as a potential tool in aerodynamic investigations of buildings. However, by considering the constraints of experimental testing in wind tunnels that limit producing the low-frequency large-scale turbulence and the inherent issues with limited integral length scale, implementing appropriate turbulence closure in CFD and developing a proper inlet transient velocity may alleviate the issues with experimental measurements in wind engineering. In CFD, the scale is not an immediate issue, as a full-scale model of the structure can be modeled and tested under various extreme wind scenarios. The simulation can be repeated to yield the same results any time. Even large-scale problems, such as simulating an urban area with condensed high-rise buildings for pollutant dispersion studies can be performed in CFD [33]; this can be challenging in laboratory testing due to scale issues.

CFD is gaining popularity within the wind engineering community along with the rise of computational power. Nowadays, CFD is commonly used to address wind engineering problems such as pollutant dispersion, wind comfort for pedestrians, aerodynamic loads on structures, or effects of bridge scour [34–37]. CFD-based numerical simulations will eventually complement the existing experimental practices for a number of wind engineering applications [38–40]. In most cases, numerical approaches are less time-consuming than experiments, and detailed information at higher resolution can be retrieved for scaled models from numerical simulations. In few earlier studies, the accuracy of analyzing bluff bodies with CFD has been questioned [41–43]. The reason behind inaccuracies was detachment of shear layer at sharp edges of bluff bodies. Detachment of shear layer makes the overall flow in the domain more responsive to local behaviors. The local effects are influenced by turbulence intensity and turbulence length scales of the incoming flow [36, 44]. Inaccurate replication of incoming turbulence properties in earlier studies was considered a reason for discrepancies in results. In Ref. [45], careful replication of horizontal turbulence properties at roof height of low-rise buildings was declared important. Few earlier studies focused on comparing surface pressures from numerical simulations with experiments and full-scale measurements. Good agreement was found among different data sources for mean pressure coefficient, while differences were found for fluctuating pressure coefficient [46].

Large eddy simulation (LES) can yield better results than turbulence closures that are based on Reynolds-Averaged Navier-Stokes (RANS), however, for higher cost of computations. The accuracy of solution of any wind engineering problem with CFD depends on the precise simulation of wind flow. A number of studies have indicated better performance of LES turbulence model for predicting mean and instantaneous flow field around bluff bodies [42, 47]. The concept of LES involves resolving the large scales in fluid flow and modeling the small scales. This approach is theoretically suitable for wind engineering applications as normally large scales are responsible for forces of interest [42]. Earlier applications of LES involving treatment of flows at low-Reynolds number yielded satisfactory results. Simply, the use of LES does not guarantee meaningful and accurate results. For flows with higher turbulence, results become more sensitive to the quality of the model [42]. Modeling of small-scale turbulence has gone through stages of improvement over the years. Sub-grid scale modeling remains the commonly used modeling technique. To yield accurate results, maintaining proper inflow boundary condition (IBC) is fundamental. Three methods are identified for generating IBC, and they are [48] (a) precursor database, (b) recycling method, and (c) turbulence synthesizing. The first two methods are computationally demanding; the third method is promising [49].

Maintaining horizontal homogeneity in the computational domain is another challenge in CFD simulations. Horizontally homogeneous boundary layer refers to the absence of artificial acceleration near the ground or stream-wise gradients in vertical profiles of mean velocity and turbulence intensity [50]. One may run steady-state simulation until it reaches convergence and monitors the vertical profiles of velocity and turbulence intensity at different locations in the domain. In case of LES, the mean value should be taken from the velocity time history for monitoring the vertical profiles. Achieving horizontal homogeneity ensures that the inlet, approach and incident flow are the same and eventually provide results with higher accuracy [50]. In several previously conducted studies, maintaining a consistent profile of mean wind speed and turbulent kinetic energy was an issue with different turbulence closure models. Significant near wall flow acceleration was found to cause unwanted change in mean wind speed and turbulent kinetic energy in simulation conducted in [51]. Additionally, issues in maintaining a consistent profile for turbulent kinetic energy were observed in [52, 53]. For accurate CFD results, maintaining consistent vertical profiles throughout the domain is important. Minor change in the profiles can create significant changes in the flow field. For flow around buildings, the importance of retaining the vertical flow profiles was stressed in Refs. [50, 54].

5. Aerodynamic testing

In Section 2, the main characteristics of ABL wind were presented. One of the main parts of any wind engineering study is to appropriately reproduce the wind characteristics in a controlled manner, to examine the response of a structure in the scope of a certain wind event. This means that first the wind flow characteristics should be simulated following an acceptable protocol and following that wind-induced pressures and loads on the surfaces of a building can be obtained by aerodynamic testing, according to the laws of similitude [55]. In order to satisfy these requirements, there are some tools used for ABL processes, including wind tunnels and open-jet facilities [56].

5.1 Wind tunnel testing

For several decades, wind tunnel modeling has been widely used as a technique to estimate wind-induced pressures and loads on buildings. **Figure 7** shows a view of a wind tunnel at the University of Western Ontario and a 1:100 scale low-rise building model. The arrangement and height of passive roughness elements are designed to reproduce wind flow over an open-terrain exposure with z_0 = 0.01 m [57]. This test case was selected benchmark for validation and comparison with other computational and experimental measurements. For accurate estimation of aerodynamic forces on buildings, proper replication of wind speed, turbulence intensity profiles, and spectral characteristics is essential [58]. Matching the spectral

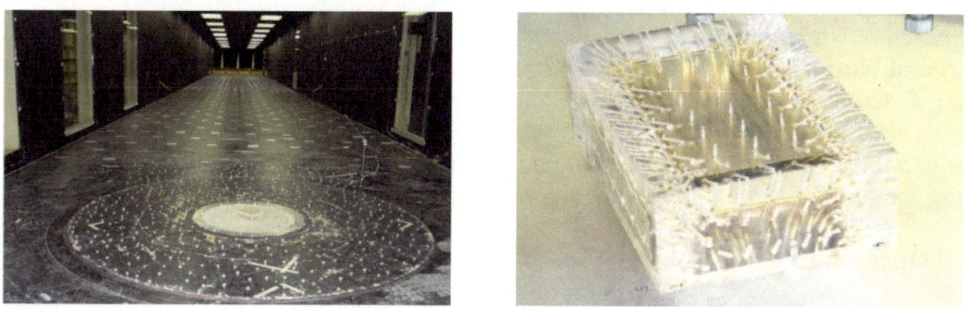

(a) (b)

Figure 7.
A view of a wind tunnel at the University of Western Ontario: (a) 1:100 low-rise building model and the roughness element arrangement for an open-terrain exposure simulating z_o = 0.01 m *and (b) a closer view of the test model instrumented with pressure taps [57].*

content of real wind flow over the entire frequency range of interest has been a major challenge in laboratory testing [30]. Duplication of the entire range of spectral content requires equality of Reynolds number. In traditional wind tunnels, small-scale turbulence can be generated. For cases where incident flow contains only small-scale turbulence, the vortices are shed downstream before attaining maturity or before creation of maximum peak pressure. The increase in large-scale turbulence content in incident flow permits vortices to attain maturity, and as a result higher peak pressures on building models are obtained [59]. The low-frequency part on the turbulence spectrum corresponds to large-scale turbulence content of the incoming flow.

The gap between small and large wavelengths of velocity fluctuations (frequency domain), for real atmospheric flows, is larger than that in wind tunnel flows. It is challenging to duplicate both small and large scales of turbulence in wind tunnels due to the absence of Reynolds number equality [59]. Moreover, the neutral atmospheric boundary layer is scaled down in the order of 1:100 to 1:500 in wind tunnels. If low-rise buildings are scaled down in a similar proportion, additional problems may be encountered. The issues with too small test models are (a) inability to modeling structural details accurately, (b) lack of aerodynamic surface pressures at higher resolution, and (c) interference effects of measuring devices [59, 60]. In practice, larger test models with scales in the order of 1:50 are used to minimize these issues. This leads to mismatch in scaling ratio of the model and the generated boundary layer, which is responsible for difference in turbulence spectra in experiments and full-scale situation. The difference in turbulence spectra is considered to be a primary reason for the large variation in aerodynamic pressures on low-rise buildings for different wind tunnel experiments [60].

Several experiments have been conducted on scaled low-rise building models and heliostats over the past few decades. Large variation in tests has been attributed to difference in Reynolds number, turbulence spectrum, geometric scaling ratio, etc. While studying the influence of turbulence characteristics on peak wind loads on heliostats, wind tunnel tests were performed, the turbulence intensity and size of the largest vortices had a noticeable effect on peak pressures, compared to other parameters Reynolds number [61]. For solar panels, peak pressures in the wind tunnel were underestimated compared to full-scale data [62]. Geometric scaling is found to be a primary source of inconsistent results in wind tunnels with similar mean flow condition [60]. It was recommended to correctly model the high-frequency end of spectrum in order to obtain acceptable mean pressure coefficients. However, for accurate mean and peak pressures, the importance of replicating the entire turbulence spectra in large-scale testing was highlighted [27]. The size of the wind tunnel was held responsible for mismatch in the low-frequency end of the spectrum. High-frequency vortices are responsible for creating the flow pattern around bluff bodies, whereas low-frequency large eddies have higher influence on aerodynamic peak loads [63]. To conclude, in the case of low-rise buildings, it has been always a challenge, in wind tunnel testing, to properly simulate wind effects due to the lack of capability in turbulence modeling [56]. As a result, other concepts and tools such as open-jet testing were devised in recent years.

5.2 Open-jet testing

As part of developing ABL simulation capabilities, a small open-jet facility was built at the Windstorm Impact, Science and Engineering (WISE) research lab, Louisiana State University (LSU) (**Figure 8**). The concept of open-jet testing is that unlike wind tunnels, the flow has no physical boundaries which has two main advantages: (i) larger eddies can be produced, leading to higher peak pressure coefficients, similar to those at full scale, and (ii) minimum blockage can be

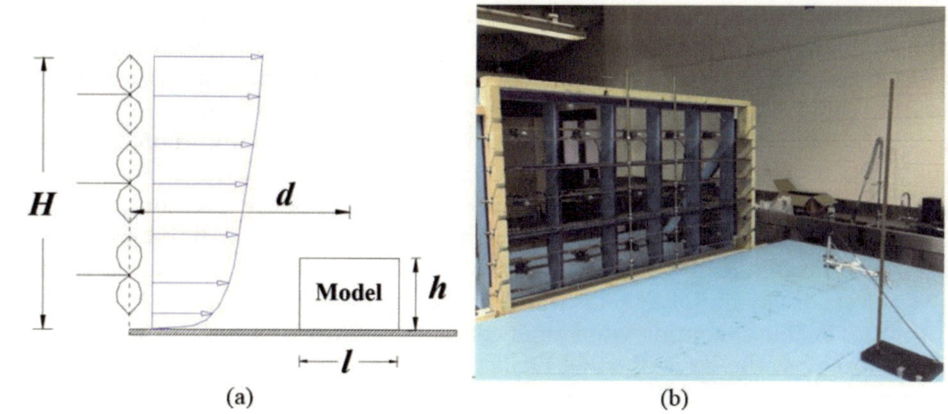

Figure 8.
The concept of open-jet testing: (a) test model located at an optimal distance from the blowers' exit and (b) 15-fan small open jet at LSU.

Figure 9.
LSU WISE small open-jet hurricane simulator (with adjustable turbulence and profile production mechanism): (1) general view of testing setup, (2) section model test specimen, (3) cobra probes for measuring 3-component wind velocities, (4) ZOC23b miniature pressure scanner, and (5) lap top computer with software for data collection and processing.

achieved. The aim was to physically simulate hurricane wind flows with similar characteristics to those of open and suburban terrain. Small-scale models of low-rise buildings were tested to examine how the turbulence structure of the approaching flow, scale issue, and open-jet exit proximity effect can influence the flow pattern on low-rise buildings and alter the separation bubble length on the roof surface. Specifically, the aim was to understand how these parameters affect the values of peak fluctuating external pressures on the roof surface [58, 64]. With an adjustable turbulence producing mechanism, different wind profiles are physically simulated. In addition, this lab has cobra probes, load cells, laser displacement sensors, and a 256-channel pressure scanning system (**Figure 9**).

A facility capable of testing low-rise buildings at full-scale would be ideal, if the artificial flow is also at full scale, which is difficult to achieve. A 1:1 scale flow that mimics real hurricane characteristics at full-scale would need giant blowers located at a distance that is significantly far than what a feasible facility can afford. Artificial wind contains significant high-frequency turbulence with limitations on the large-size vortices that make scaling buildings unavoidable, if we were to replicate correct physics. There are some testing capabilities that can engulf full-scale residential homes; however, the flow characteristics raises important questions about their similarity to those at full scale. This said, scaling residential homes is essential to maintain correct physics, and at the same time large-scale testing (not full-scale) will lead to improved Reynolds number. Large-scale wind testing went through several phases before reaching the present stage [63]. A multidisciplinary LSU research team from Civil and Environmental Engineering, Mechanical Engineering, Coast and Environment, Louisiana Sea Grant, Geography and Anthropology, Construction Management, and Sociology collaborated on a project titled "Hurricane Flow Generation at High Reynolds Number for Testing Energy and Coastal Infrastructure" that was awarded by the Louisiana Board of Regents to build Phase 1 of a large wind and rain testing facility (**Figure 10**). Phase 1 permits generating wind flows at a relatively high Reynolds number over a test section of 4 m × 4 m. These capabilities enable executing wind engineering experiments at relatively large scales. Moreover, the large open-jet facility has a potential for conducting destructive testing on models built from true construction materials. Blockage is minim, as per the concept of open-jet testing [65]. This state-of-the-art facility can generate realistic hurricane wind turbulence by replicating the entire frequency range of the velocity spectrum.

The large LSU WISE open-jet facility enables researchers to test their research ideas; to expand knowledge leading to innovations and discovery in science, hurricane engineering, and materials and structure disciplines; and to build the more resilient and sustainable infrastructure. The facility will enable scientists and researchers to test potential mitigation and restoration solutions, both natural (e.g., vegetation) and artificial. Potential applications include, but are not limited to, wind turbines, solar panels, residential homes, large roofs, high-rise buildings, transportation infrastructure, power transmission lines, etc. Testing at this facility can provide knowledge useful for homeowners and insurance companies to deal more effectively with windstorms, for example, to fine tune design codes and give coastal residents options for making their dwellings more storm-resistant. The goal is to build new structures and retrofit existing ones in innovative ways to balance resilience with sustainability, to better protect people, to enhance safety, and to reduce the huge cost of rebuilding after windstorms. In addition, the facility offers tremendous education value to k-12, undergraduate, and graduate students at a flagship state university, designated as a land-grant, sea-grant, and space-grant institution. This will broadly impact the wind/structural engineering research and

Figure 10.
LSU WISE large testing facility (with a test section of 4 m × 4 m).

education field and facilitate effective investments in the infrastructure industry that will result in more resilient and sustainable communities and contribute to economic growth and improve the quality of life.

6. Sample study of building aerodynamics at the LSU WISE lab

The LSU research team aspires to match the spectral content of real wind using large-scale open-jet testing and CFD simulations in their quest of accurate estimation of peak pressures on building surfaces under wind. The goal is precise estimation of peak pressures on buildings through the generation of large- and small-scale turbulence via open-jet testing as well as advanced CFD simulations. Extreme negative pressures near ridges, corners, and leading edges of roofs are governed dominantly by wind turbulence and Reynolds number, among other factors. Both small- and large-scale turbulence vortices are responsible for peak pressures and can influence separation in the shear layer. This demands for precise replication of wind speed profile, turbulence intensity profiles, and spectral characteristics. Replication of the true physics requires higher Reynolds number which is difficult to achieve in wind tunnels. In traditional wind testing, it is challenging to create large-scale turbulence. An increase in large-scale turbulence content in incident flow allows vortices to attain maturity, and as a result higher peak pressures can be reproduced. A fundamental research objective, however, is to address the challenge of replicating real wind turbulence experimentally and computationally. Resolving the scaling issue by investigating larger test models at higher Reynolds number is another highlight of our research at the LSU WISE lab.

The velocity was measured at different heights in the open-jet facility with cobra probes to obtain the mean velocity and turbulence intensity profiles. **Figure 11(i)** shows the comparison of experimental mean velocity profiles from LSU open jet and TPU wind tunnel with theoretical wind profiles measured for open terrain condition ($z_0 = 0.01\,m$). It was observed that the measured mean velocity profile at LSU open jet was consistent with different theoretical profiles and also velocity profile measured at TPU. It should be noted that the U_{ref} corresponding to 10 m in full scale was 22 m/s [66]. For normalizing the experimental data, velocity information corresponding to $H_{ref} = 0.75$ m was considered. Mean velocity corresponding to 0.75 m was considered to be U_{ref} in the open jet. **Figure 11(ii)** shows along wind turbulence intensity profiles from experimental data and theoretical formulations. The velocity data were processed to obtain turbulence intensity, and the profile was compared with theoretical profile corresponding to the following equation.

$$I_u(z) = \frac{1}{\ln\left(\frac{z}{z_0}\right)} \tag{10}$$

The vertical profile plot for turbulence intensities shows that the LSU open-jet facility has approximately 20% turbulence intensity at reference height. Both mean velocity and turbulence intensity profiles in **Figure 11** shows that LSU open-jet facility is capable of replicating open terrain near-ground ABL flow.

A scaled (1:13) cubic building model was tested at the LSU WISE large open-jet facility. The primary objective of this task was to compare surface pressure coefficients those obtained by wind tunnel testing on a smaller scale (1:100) model. Wind tunnel measurements are obtained from the published dataset of Tokyo Polytechnic University (TPU). The building model was instrumented with several pressure taps to capture surface pressures. A total of 64 taps were distributed on roof, same as the

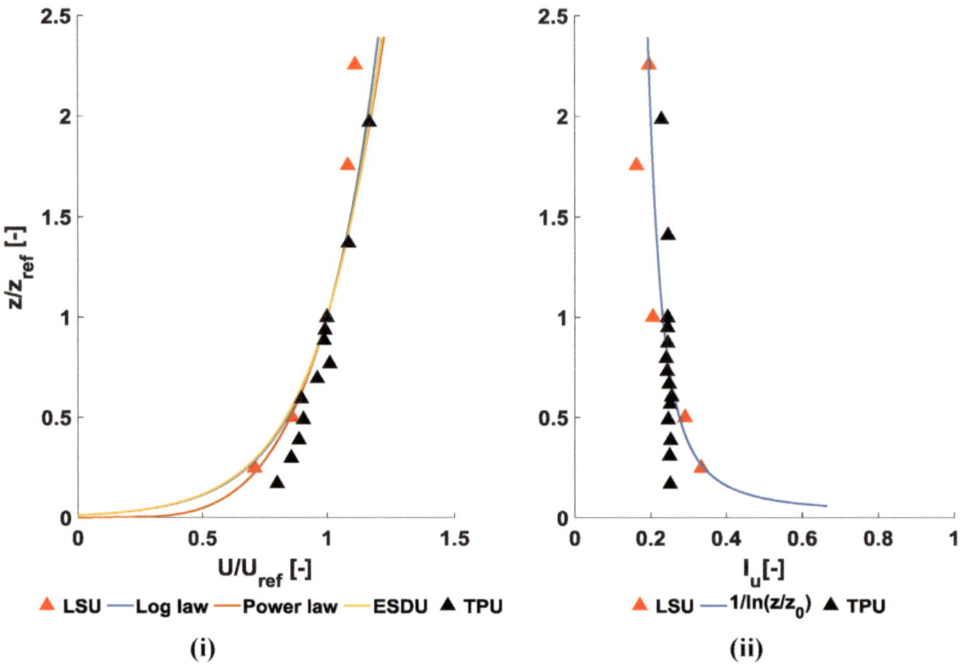

Figure 11.
Flow characteristics: (i) mean velocity and (ii) turbulence intensity.

TPU wind tunnel model. Pressure taps were connected to Scanivalve pressure scanners through appropriate tubing. Two cobra probes were used to monitor upstream velocity at roof height [58]. The following equation was used to compute the pressure coefficient.

$$C_p(t) = \frac{p(t) - p_s}{\left(\frac{1}{2}\right)\rho U^2} \tag{11}$$

The time history of pressure coefficients, $C_p(t)$, was obtained from the pressure time history, $p(t)$, recorded using pressure scanners. The static pressure p_s was considered reference for the pressure transducers. In addition, base line pressures were collected before and after each experiment. Once the time history of pressure coefficients was obtained, statistical analysis was done to obtain mean, minimum, and root mean square (rms) values. Measurements from LSU open-jet and TPU wind tunnel were processed the same way. The maximum and minimum values were obtained using MATLAB functions with a probabilistic approach described in Ref. [67]. This approach was considered, to account for the highly fluctuating wind flow, to yield a more stable estimator of peak values.

Sample of the findings of the experiment and comparison with TPU results is shown in **Figure 12**. The distribution of pressure coefficients obtained by open jet testing is symmetric like what is observed in the TPU wind tunnel testing. Since the model in open jet was tested at a higher Reynolds number, higher values of peak pressure coefficients are realized. Higher suction was observed near the zone of flow separation on the roof. Stronger suction for open-jet testing was found due to higher Reynolds number in open-jet and the presence of larger-scale turbulence compared to the wind tunnel. This difference in Reynolds number leads to difference in formation of flow separation zone, stagnation point on windward face, and the reattachment length. The difference between full-scale and reduced-scale wind tunnel tests is owed to similar reasons.

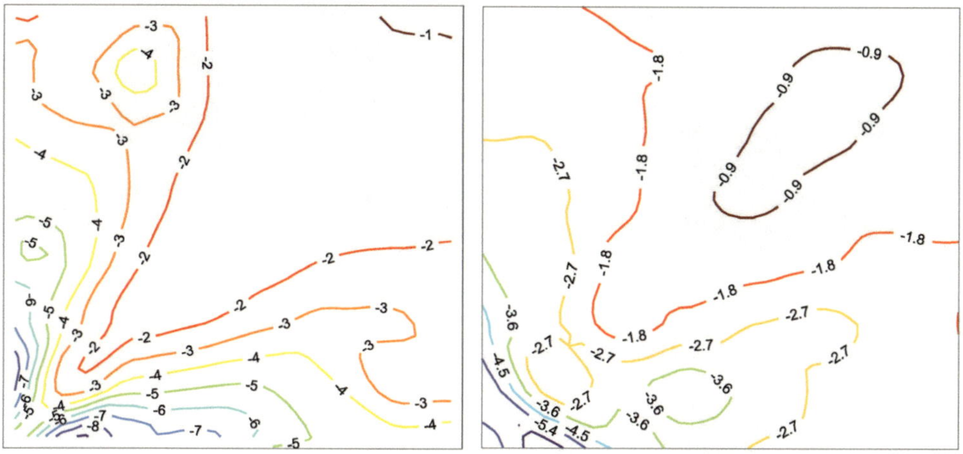

(a) (b)

Figure 12.
Minimum pressure coefficients on roof: (a) LSU open-jet (b) TPU wind tunnel (wind from bottom left corner).

On the computational side, the k- SST turbulence model was employed for improved mean pressure prediction near the flow separated region. An advanced approach is ongoing that employs large eddy simulation (LES) to generate accurate mean and peak pressures. **Figure 13** shows a sample of high-quality mesh and CFD simulations in OpenFOAM.

In order to alleviate the challenges and shortcomings involved within the experimental tests in boundary-layer wind tunnels, in recent years, CFD was considered as an effective tool for the simulation of wind effects on civil engineering structures. However, it is necessary that the numerical CFD model would be capable of generating turbulence in a flow with certain spectral contents and eventually to reproduce peak pressures on building surfaces. The objective of this research is therefore to provide a basis for the development of recommendations and guidelines on using a CFD LES model that enables appropriate simulation of turbulence spectra of ABL inflow and reproducing the peak wind pressures on the roof of low-rise buildings. **Figure 14** represents a schematic of the tools used by Aly and Gol Zaroudi [49] to simulate peak wind loads on a benchmark full-scale building from the Texas Tech University (TTU) in an open-terrain field. The details, advantages, and disadvantages of each tool are discussed in Aly and Gol Zaroudi [49].

Considering the current rapid improvements in developing high-speed processors that can run in parallel on high-performance computing (HPC) clusters and devising new digital storage devices with huge capacities, CFD is becoming a promising tool in wind engineering applications. However, it is still a challenge for proper simulation of turbulence according to ABL wind characteristics and accurately reproducing peak pressures on low-rise buildings, even with supercomputers [40]. Aly and Gol Zaroudi [49], therefore, attempt to address some of the challenges in experimental and numerical simulations for aerodynamic testing of low-rise buildings, to reproduce realistic peak pressures. The study focused on wind flow processes in CFD with an objective to mimic full-scale pressures on low-rise building. The study implemented CFD with LES on a scale of 1:1 building. After a proximity experiment was executed in CFD-LES, a location of the test building from the inflow boundary was recommended, different from existing guidelines (RANS-based, e.g., COST and AIJ). The inflow boundary proximity showed significant influence on pressure correlation and the reproduction of peak pressures.

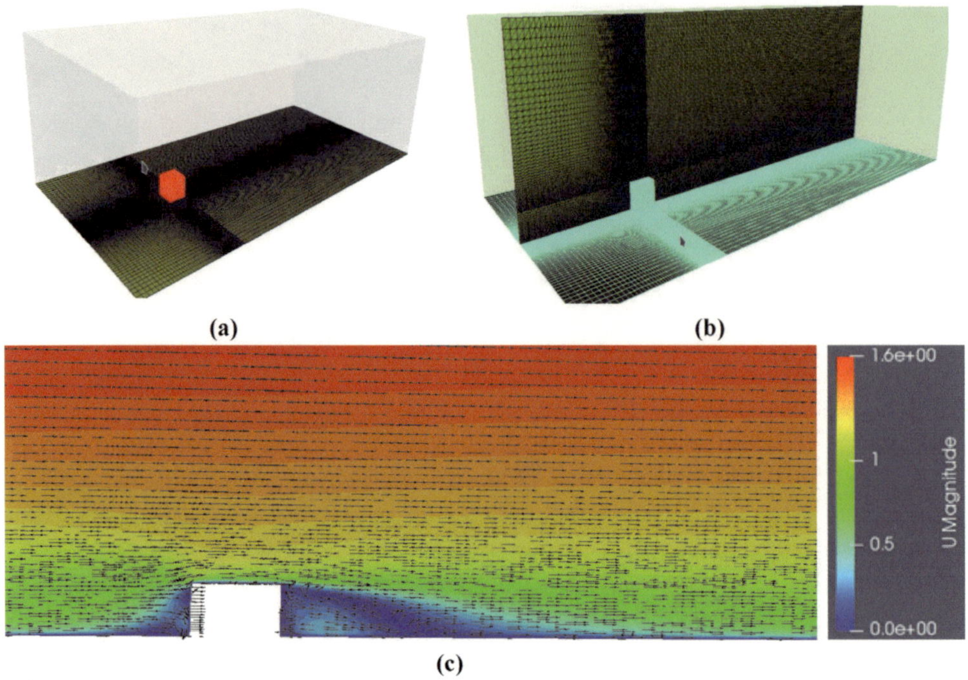

Figure 13.
With high-quality mesh and potential turbulence closure, CFD can provide continuous flow information: (a) 3D view of the computational grid, (b) meshing arrangement along the longitudinal section over a cube, and (c) velocity contour, after simulations in OpenFOAM.

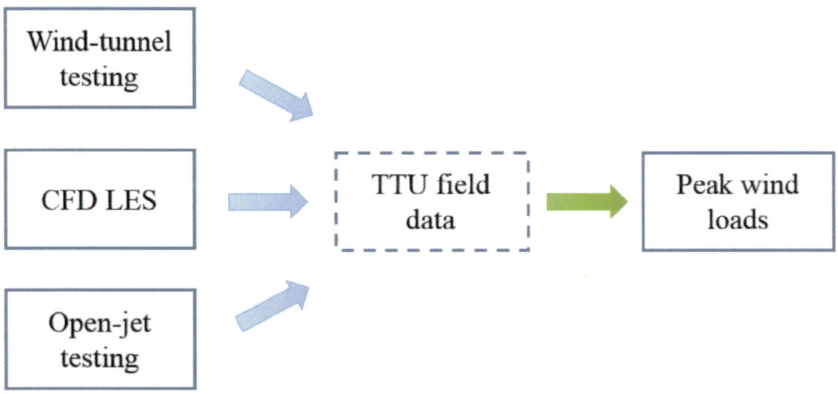

Figure 14.
The research tools employed to reproduce peak wind pressures on the roof of a benchmark low-rise building from the Texas Tech University (TTU) in an open-terrain field.

The CFD LES turbulence closure showed its capabilities to reproduce peak loads that can mimic field data owing to the ability of creating inflow with enhanced spectral contents at 1:1 scale [49].

7. Concluding remarks

This chapter described the main characteristics of ABL winds, as well as some available tools for aerodynamic testing. Earlier studies confirm the presence of extreme negative pressures near ridges, corners, and leading edges of roofs in wind events. Turbulence (small- and large-scale) is responsible for large peak negative pressures and separation in the shear layer. This demands for precise replication of wind speed profile, turbulence intensity profiles, and spectral characteristics. Replication of true physics requires equality of Reynolds number which is not possible in wind tunnels. In traditional wind tunnels, only small-scale turbulence can be generated. An increase in large-scale turbulence content in incident flow allows vortices to attain maturity, and as a result higher peak pressures are obtained. The challenge of properly simulating wind effects on low-rise buildings is related to the lack of capability in turbulence modeling at a reasonably large scale and its limitation in reproducing the low-frequency part of the ABL turbulence spectrum. As a result, advances in aerodynamic testing employing modern tools such as open-jet testing for large- and full-scale testing were devised in recent years. Resolving the scaling issue by studying larger models at higher Reynolds number is another highlight of recent advances in aerodynamic testing. A large-scale cubic building model was tested in LSU open-jet facility at higher Reynolds number, and pressure coefficients were compared with those from wind tunnel testing. The results reveal the importance of large-scale testing at higher Reynolds numbers to obtain realistic peak pressures. Furthermore, CFD with appropriate turbulence closure was widely implemented recently for full-scale studies of wind effects on civil engineering structures. However, adopting proper inlet transient velocity is very crucial to correctly simulate ABL wind characteristics.

Author details

Aly Mousaad Aly*, Faiaz Khaled and Hamzeh Gol-Zaroudi
Windstorm Impact, Science and Engineering (WISE) Laboratory, Louisiana State University, Baton Rouge, LA, United States

*Address all correspondence to: aly@LSU.edu

References

[1] Knutson TR, McBride JL, Chan J, Emanuel K, Holland G, Landsea C, et al. Tropical cyclones and climate change. Nature Geoscience. 2010;**3**: 157-163

[2] Mann ME, Emanuel KA. Atlantic hurricane trends linked to climate change. Eos, Transactions American Geophysical Union. 2006;**87**: 233-241

[3] Walsh KJE, McBride JL, Klotzbach PJ, Balachandran S, Camargo SJ, Holland G, et al. Tropical cyclones and climate change. WIREs Climate Change. 2016;7:65-89. DOI: 10.1002/wcc.371

[4] Pielke RA Jr, Gratz J, Landsea CW, Collins D, Saunders MA, Musulin R. Normalized hurricane damage in the United States: 1900–2005. Natural Hazards Review. 2008;**9**:29-42

[5] U S Federal Emergency Management Agency. 2017 Hurricane Season, FEMA After Action Reports 2018. Available from: https://www.fema.gov/media-library-data/1531743865541-d16794d43d3082544435e1471da07880/2017FEMAHurricaneAAR.pdf

[6] ASCE7-2010. Minimum design loads for buildings and other structures, ASCE Standard, ASCE/SEI 7-10. Reston, Virginia: American Society of Civil Engineers; 2010

[7] He J, Pan F, Cai C. A review of wood-frame low-rise building performance study under hurricane winds. Engineering Structures. 2017;**141**: 512-529

[8] New York Daily News. Hurricane Sandy Damages Pictures; 2012

[9] WingsoflLove. Katrina Rita Pics 2020. Available from: http://www.wingsoflove.org/KatrinaRitaPics.htm

[10] Crandell JH, Nowak M, Laatsch EM, van Overeem A, Barbour C, Dewey R, et al. Assessment of Damage to Single-Family Homes Caused by Hurricanes Andrew and Iniki. Washington, DC: United States Department of Housing and Urban Development, Office of Policy Development and Research; 1993

[11] WPC. Hurricane Charley 2004 Damage Survey. Bermuda: Wyndham Partners Consulting, Limited; 2004

[12] Baskaran A, Molleti S, Roodvoets D. Understanding low-sloped roofs under hurricane Charley from field to practice. Journal of ASTM International. 2007;**4**: 1-13. DOI: 10.1520/JAI101055

[13] Holmes JD. Wind Loading of Structures. 3rd ed. Boca Raton, Florida: CRC Press; 2015. Available from: https://www.crcpress.com/Wind-Loading-of-Structures/Holmes/p/book/9781138893979

[14] Simiu E. Design of Buildings for Wind – A Guide for ASCE 7–10 Standard Users and Designers of Special Structures. 2nd ed. NY: John Wiley & Sons, Inc.; 2011

[15] FloridaDisaster. Florida division of emergency management. 2020. Avilable from: https://www.floridadisaster.org/

[16] Isyumov N, Alan G. Davenport's mark on wind engineering. Journal of Wind Engineering and Industrial Aerodynamics. 2012;**104**:12-24. DOI: 10.1016/j.jweia.2012.02.007

[17] Kaimal JC, Wyngaard JC, Izumi Y, Coté OR. Spectral characteristics of surface-layer turbulence. Quarterly Journal of the Royal Meteorological Society. 1972;**98**:563-589. DOI: 10.1002/qj.49709841707

[18] Simiu E, Scanlan RH. Wind Effects on Structures: Fundamentals and

Applications to Design. New York: Wiley; 1996

[19] ESDU 85020. Characteristics of atmospheric turbulence near the ground. Part II: single point data for strong winds (neutral atmosphere). Eng Sci Data Unit, Data Items; 2001

[20] Fujimura M, Maeda J. Cross-correlation of fluctuating components of wind speed based on strong wind measurement. In: Proceedings of the 7th Pacific-Asia Conference on Wind Engineering. Taipei: T1-D; 2009

[21] Maeda J, Makino M. Classification of customary proposed equations related to the component of the mean wind direction in the structure of atmospheric turbulence and these fundamental properties. Transactions of the Architectural Institute of Japan. 1980; **287**:77-87

[22] Roach P. The generation of nearly isotropic turbulence by means of grids. International Journal of Heat and Fluid Flow. 1987;**8**:82-92. DOI: 10.1016/0142-727X(87)90001-4

[23] El-Gabry L, Thurman D, Poinsatte P. Procedure for Determining Turbulence Length Scales Using Hotwire Anemometry. OH: Cleveland; 2014 https://ntrs.nasa.gov/search.jsp?R=20150000733

[24] Solari G. Gust buffeting. I: Peak wind velocity and equivalent pressure. Journal of Structural Engineering. 1993; **119**:365-382

[25] Solari G. Gust buffeting. II: Dynamic alongwind response. Journal of Structural Engineering. 1993;**119**: 383-398

[26] Holmes JD, Sankaran R, Kwok KCS, Syme MJ. Eigenvector modes of fluctuating pressures on low-rise building models. Journal of Wind Engineering and Industrial Aerodynamics. 1997;**69–71**:697-707. DOI: 10.1016/S0167-6105(97)00198-0

[27] Richards PJ, Hoxey R, Connell BD, Lander DP. Wind-tunnel modelling of the Silsoe cube. Journal of Wind Engineering and Industrial Aerodynamics. 2007;**95**:1384-1399. DOI: 10.1016/j.jweia.2007.02.005

[28] Fu TC, Aly AM, Chowdhury AG, Bitsuamlak G, Yeo DH, Simiu E. A proposed technique for determining aerodynamic pressures on residential homes. Wind and Structures An International Journal. 2012;**15**:27-41. DOI: 10.12989/was.2012.15.1.027

[29] Simiu E. Toward a standard on the wind tunnel method. NIST Technical Note. 2009;**1655**:1-36

[30] Aly AM. On the evaluation of wind loads on solar panels: The scale issue. Solar Energy. 2016;**135**:423-434. DOI: 10.1016/j.solener.2016.06.018

[31] Ho TCE, Surry D, Morrish D, Kopp GA. The UWO contribution to the NIST aerodynamic database for wind loads on low buildings: Part 1. Archiving format and basic aerodynamic data. Journal of Wind Engineering and Industrial Aerodynamics. 2005;**93**:1-30. DOI: 10.1016/j.jweia.2004.07.006

[32] Sagaut P, Deck S, Terracol M. Multiscale and Multiresolution Approaches in Turbulence. 2nd ed. London: World Scientific; 2013. DOI: 10.1142/p878

[33] Gousseau P, Blocken B, Stathopoulos T, van Heijst GJF. CFD simulation of near-field pollutant dispersion on a high-resolution grid: A case study by LES and RANS for a building group in downtown Montreal. Atmospheric Environment. 2011;**45**: 428-438. DOI: 10.1016/j.atmosenv.2010.09.065

[34] Yoshie R, Mochida A, Tominaga Y, Kataoka H, Harimoto K, Nozu T, et al.

Cooperative project for CFD prediction of pedestrian wind environment in the architectural Institute of Japan. Journal of Wind Engineering and Industrial Aerodynamics. 2007;**95**:1551-1578. DOI: 10.1016/j.jweia.2007.02.023

[35] Nozu T, Tamura T, Okuda Y, Sanada S. LES of the flow and building wall pressures in the center of Tokyo. Journal of Wind Engineering and Industrial Aerodynamics. 2008;**96**:1762-1773. DOI: 10.1016/j.jweia.2008.02.028

[36] Nozawa K, Tamura T. Large eddy simulation of the flow around a low-rise building immersed in a rough-wall turbulent boundary layer. Journal of Wind Engineering and Industrial Aerodynamics. 2002;**90**:1151-1162. DOI: 10.1016/S0167-6105(02)00228-3

[37] Xiong W, Cai CS, Kong B, Kong X. CFD simulations and analyses for bridge-scour development using a dynamic-mesh updating technique. Journal of Computing in Civil Engineering. 2016;**30**(1):04014121. DOI: 10.1061/(ASCE)CP.1943-5487.0000458

[38] Murakami S. Current status and future trends in computational wind engineering. Journal of Wind Engineering and Industrial Aerodynamics. 1997;**67–68**:3-34. DOI: 10.1016/S0167-6105(97)00230-4

[39] Blocken B. 50 years of computational wind engineering: Past, present and future. Journal of Wind Engineering and Industrial Aerodynamics. 2014;**129**:69-102. DOI: 10.1016/j.jweia.2014.03.008

[40] Cochran L, Derickson R. A physical modeler's view of computational wind engineering. Journal of Wind Engineering and Industrial Aerodynamics. 2011;**99**:139-153. DOI: 10.1016/j.jweia.2011.01.015

[41] Rodi W. Large-eddy simulations of the flow past bluff bodies: State-of-the art. JSME International Journal Series B: Fluids and Thermal Engineering. 1998;**41**(2):361-374. DOI: 10.1299/jsmeb.41.361

[42] Shah KB, Ferziger JH. A fluid mechanicians view of wind engineering: Large eddy simulation of flow past a cubic obstacle. Journal of Wind Engineering and Industrial Aerodynamics. 1997;**67–68**:211-224. DOI: 10.1016/S0167-6105(97)00074-3

[43] Lim HC, Thomas TG, Castro IP. Flow around a cube in a turbulent boundary layer: LES and experiment. Journal of Wind Engineering and Industrial Aerodynamics. 2009;**97**(2):96-109. DOI: 10.1016/j.jweia.2009.01.001

[44] Ricci M, Patruno L, de Miranda S. Wind loads and structural response: Benchmarking LES on a low-rise building. Engineering Structures. 2017;**144**:26-42. DOI: 10.1016/j.engstruct.2017.04.027

[45] Tieleman HW, Hajj MR, Reinhold TA. Wind tunnel simulation requirements to assess wind loads on low-rise buildings. Journal of Wind Engineering and Industrial Aerodynamics. 1998;**76**:675-685

[46] Richards PJ, Wanigaratne BS. A comparison of computer and wind-tunnel models of turbulence around the Silsoe structures building. Journal of Wind Engineering and Industrial Aerodynamics. 1993;**46–47**:439-447. DOI: 10.1016/0167-6105(93)90310-K

[47] Rodi W. Comparison of LES and RANS calculations of the flow around bluff bodies. Journal of Wind Engineering and Industrial Aerodynamics. 1997;**69–71**:55-75. DOI: 10.1016/S0167-6105(97)00147-5

[48] Keating A, Piomelli U, Balaras E, Kaltenbach HJ. A priori and a posteriori tests of inflow conditions for large-eddy

simulation. Physics of Fluids. 2004;**16**: 4696-4712. DOI: 10.1063/1.1811672

[49] Aly AM, Gol-Zaroudi H. Peak pressures on low rise buildings: CFD with LES versus full scale and wind tunnel measurements. Wind and Structures. 2020;**30**:99. DOI: 10.12989/was.2020.30.1.099

[50] Blocken B, Stathopoulos T, Carmeliet J. CFD simulation of the atmospheric boundary layer: Wall function problems. Atmospheric Environment. 2007;**41**:238-252. DOI: 10.1016/j.atmosenv.2006.08.019

[51] Zhang CX. Numerical predictions of turbulent recirculating flow with a k-epsilon model. Journal of Wind Engineering and Industrial Aerodynamics. 1994;**51**:177-201

[52] Quinn AD, Wilson M, Reynolds AM, Couling SB, Hoxey RP. Modelling the dispersion of aerial pollutants from agricultural buildings - an evaluation of computational fluid dynamics (CFD). Computers and Electronics in Agriculture. 2001;**30** (1–3):219-235. DOI: 10.1016/S0168-1699 (00)00166-6

[53] Riddle A, Carruthers D, Sharpe A, McHugh C, Stocker J. Comparisons between FLUENT and ADMS for atmospheric dispersion modelling. Atmospheric Environment. 2004;**38**(7): 1029-1038. DOI: 10.1016/j.atmosenv.2003.10.052

[54] Blocken B, Carmeliet J, Stathopoulos T. CFD evaluation of wind speed conditions in passages between parallel buildings-effect of wall-function roughness modifications for the atmospheric boundary layer flow. Journal of Wind Engineering and Industrial Aerodynamics. 2007;**95**(9–11):941-962. DOI: 10.1016/j.jweia.2007.01.013

[55] ASCE. 49–12. Wind Tunnel Testing for Buildings and Other Structures. American Society of Civil Engineers: Reston, VA; 2012. DOI: 10.1061/9780784412282

[56] Aly AM. Atmospheric boundary-layer simulation for the built environment: Past, present and future. Building and Environment. 2014;**75**: 206-221. DOI: 10.1016/j.buildenv.2014.02.004

[57] Ho TCE, Surry D, Morrish DP. NIST/TTU Cooperative Agreement–Windstorm Mitigation Initiative: Wind Tunnel Experiments on Generic Low Buildings. London, Canada: BLWT-SS20–2003, Boundary-Layer Wind Tunnel Laboratory, Univ. of Western Ontario; 2003

[58] Gol-Zaroudi H, Aly AM. Open-jet boundary-layer processes for aerodynamic testing of low-rise buildings. Wind and Structures. 2017; **25**:233-259. DOI: 10.12989/was.2017.25.3.233

[59] Tieleman HW. Wind tunnel simulation of wind loading on low-rise structures: A review. Journal of Wind Engineering and Industrial Aerodynamics. 2003;**91**:1627-1649. DOI: 10.1016/j.jweia.2003.09.021

[60] Jafari A, Ghanadi F, Emes MJ, Arjomandi M, Cazzolato BS. Measurement of unsteady wind loads in a wind tunnel: Scaling of turbulence spectra. Journal of Wind Engineering and Industrial Aerodynamics. 2019;**193**: 103955. DOI: 10.1016/j.jweia.2019.103955

[61] Emes MJ, Arjomandi M, Ghanadi F, Kelso RM. Effect of turbulence characteristics in the atmospheric surface layer on the peak wind loads on heliostats in stow position. Solar Energy. 2017;**157**:284-297. DOI: 10.1016/j.solener.2017.08.031

[62] Stathopoulos T, Zisis I, Xypnitou E. Wind loads on solar collectors: A

review. Structures Congress. 2012;**2012**: 1169-1179

[63] Aly AM, Gol-Zaroudi H. Atmospheric boundary layer simulation in a new open-jet facility at LSU: CFD and experimental investigations. Measurement. 2017;**110**:121-133. DOI: 10.1016/j.measurement.2017.06.027

[64] Aly AM. Wind engineering science for resilient and sustainable structures. Louisiana Civil Engineer – November 2016. American Society of Civil Engineers. 2016;**25**(1):16-19

[65] Aly AM, Chowdhury AG, Bitsuamlak G. Wind profile management and blockage assessment for a new 12-fan wall of wind facility at FIU. Wind and Structures An International Journal. 2011;**14**:285-300

[66] Tokyo Polytechnic University. Aerodynamic database for low-rise buildings; 2007

[67] Sadek F, Simiu E. Peak non-Gaussian wind effects for database-assisted low-rise building design. Journal of Engineering Mechanics. 2002; **128**:530-539. DOI: 10.1061/(asce) 0733-9399(2002)128:5(530)

Chapter 3

Research on Aeroelasticity Phenomenon in Aeronautical Engineering

Hoang Thi Kim Dung and Nguyen Phu Khanh

Abstract

Aeroelasticity phenomena arise when structural deformations induce changes on aerodynamic forces due to airplane structures that are not completely rigid. The additional aerodynamic forces cause an increase in the structural deformations, which leads to greater aerodynamic forces in a feedback process. These interactions may become smaller until reaching a condition of equilibrium or may diverge catastrophically if resonance occurs. Flutter is an instability aeroelasticity phenomenon which is the most difficult to predict. In this chapter, a numerical method and an experimental method were realized to predict aeroelastic response and characteristic parameters of a wing structure. The numerical method was firstly developed based on the interaction between computational fluid dynamic and computational structural dynamic methods using a coupling system, fluid–solid interaction (FSI), in the ANSYS software. Then, an experiment was set up in suitable conditions to study aeroelasticity characteristics with the goal of comparing the numerical results with the experimental results on the same wing structure at low speed. After that, a developed code based on immersed boundary method (IBM) was realized to predict aeroelasticity response and characteristic parameters of the wing structure. AGARD 445.6 wing model was chosen for this developed procedure at high speed. Obtained results were compared to other numerical and experimental results.

Keywords: aeroelasticity, flutter, FSI, IBM, ANSYS

1. Introduction

Flutter is defined as the dynamic instability of aeroelasticity. Flutter is one of the most dangerous aeroelasticity phenomena as it could lead to a destroyed structure. The reason is the unsteady aerodynamic forces generated from elastic deformations of the structure that are usually involved with complicated phenomena such as the shock wave/boundary layer interaction, flow separation, nonlinear limited cycle oscillation, and more. Flutter is determined as a critical issue determining the reliability of the airplane wings or aircraft engine turbo-machine blades. Therefore, in the early phase of the structural design of the air vehicle, aircraft engine turbomachinery, flutter problems should be calculate and predicted. However, accurate prediction of the flutter is very challenging due to the perplexing physical phenomena and the required large amount of computation [1–4].

Coupled aeroelastic solution procedures use strongly coupled algorithms which contained sufficient interaction between computational fluid dynamics (CFD) and computational structural dynamics (CSD). As computer technology progresses, higher-order methods of CFD based on the Euler and the Navier-Stokes equations become more attractive due to the ability of the model and its more accurately transonic, nonlinear, and viscous effects. CFD has also advanced from two-dimensional problems to fully three-dimensional problems with or without coupled solution of the structural equations (CSD). The flow solvers used in aeroelastic analysis include 3D Euler and Navier-Stokes solvers which assumed inviscid flow.

The dynamic response of flutter characteristics of the first AGARD 445.6 wing standard aeroelastic configuration was studied using an unsteady Navier-Stokes algorithm in order to investigate a previously noted discrepancy between Euler flutter characteristics and the experimental data [5]. The 3D implicit upwind Euler/Navier-Stokes code (CFL3D Version 2.1) was previously modified for the time-marching aeroelastic analysis of wings using the unsteady Euler equations. A linear stability analysis and a time-marching aeroelastic analysis were used to determine the flutter characteristics of the isolated 45° swept-back wing. The flutter characteristics of the wing were determined using traditional V-g analysis. This stability analysis was determined at free-stream Mach numbers of 0.96 and 1.141 using the generalized aerodynamic forces calculated by solving the Euler equations and the Navier-Stokes equations.

Computational flutter required a fluid-structure interface as a common boundary to exchange the aerodynamic loads and structural displacements at the wing surface. But, the aerodynamic and structural grids were not coincident due to different systems used (fluent solver for aerodynamic grids and mechanical ADPL solver for structural grids). Therefore, the system coupling tool in ANSYS Workbench was used to transfer the aerodynamic pressure loads from the CFD grid points to the CSD grid points and vice versa, which ensured a conservative transfer of energy between the two systems [6].

Time-accurate aeroelastic simulations were carried out using the modal coupled aeroelastic implementation for a standard experimental test case: the AGARD 445.6 aeroelastic wind tunnel model in the subsonic and transonic regions [7]. A numerical methodology coupling Navier-Stokes equations and structural modal equations for predicting 3D transonic wing flutter was described in [8]. A modal approach is used for the structural response. The results indicate that the first five modes are sufficient to accurately model the wing structure response. In Ref. [9], an unsteady Reynolds-averaged Navier-Stokes (RANS) model was coupled with normal modes of structure to predict the flutter boundary for the AGARD 445.6 wing. A new integrated CFD-CSD simulation for flutter calculations based on a parallel, multiblock, multigrid flow solver for the Euler/Navier-Stokes equations using the ANSYS software was found in [10]. Computations were performed for a three-dimensional test case of AGARD 445.6 wing to validate and establish the usefulness of the simulation. Immersed boundary method solved Navier-Stokes equations for flow in couple with the Newton equation for structure movement under the effect of friction force exerted on the structure surface to carry out fluid–structure interaction (FSI). However, computational grids needed to be re-meshed in each time step due to changes of the structure position in time. To overcome this obstacle, immersed boundary method and finite volume methods were both invoked in solving the interaction between fluid flow and moving structure [11]. This research estimated the numerical and experimental results on the wing structure at a low speed with four different wing models such as two rectangular and two trapezoid 3D-shape wings, each 3D-shape wing had symmetric and asymmetric airfoil, respectively.

The subsonic aeroelastic stability of a two-dimensional panel resting on a continuous elastic foundation was investigated in Ref. [12]. Tests were conducted experimentally on a 104 × 24 × 0.018 in. rectangular aluminum panel in a low-speed wind tunnel. Comparison of experiment and theory showed a good agreement in flutter speed and wavelength but poor agreement in wave speed and frequency at flutter. This discrepancy was attributed to the limitations in the test setup as well as to the general difficulty of predicting the wave speed and frequency as accurately as the flutter speed. Reference [13] tested the first AGARD standard aeroelastic configuration for dynamic response, 445.6 wing, in the 16 foot transonic dynamics tunnel at the National Aeronautics and Space Administration (NASA) Langley Research Center. Several models of the wing were tested in the transonic dynamic tunnel including full-span and semispan models over a range of Mach number from 0.338 to 1.141. The NASA conducted experiments in wind tunnel to estimate the aeroelastic characteristics of new and advanced flight vehicles, including fixed-wing, rotary-wing, and space-launch configurations. Reviews and assessments were made regarding available facilities, measurement techniques, and other means and devices useful in testing. The needs and requirements for advances and improvements in testing capabilities for future experimental research and development programs are described [14].

In [15], aeroelastic concepts for increased aircraft performance were mentioned. Active aeroelastic concepts as well as robust analysis concepts aiming at efficient analysis were carried out using numerical models with uncertain or varying model parameters. A high aspect ratio wing in wind tunnel testing conditions was considered for exploitation of fluid-structure interaction of active aeroelastic structures. The structural flexibility was exploited by using multiple control surfaces such that the deformed wing shape gives minimum drag for different flight conditions. Two different drag minimization methods were carried out: one was to reduce induced drag based on numerical optimization techniques, and another was to reduce measured total drag using real-time optimization in the wind tunnel experiment.

An approach for the prediction of dynamic modal transient response and flutter characteristics of structures with unknown system parameters, such as stiffness and mass, using experimental modal parameters was remarked in [16]. A finite element model was created by using the actual material properties of the structure to study the correlation of the results. The computed transient responses and flutter velocities by the proposed method using experimental modal parameters observed that material properties were not a prerequisite.

In Ref. [17], transonic flutter characteristics of AGARD 445.6 wing between the numerical method [5] and the experimental data of NASA's experiment were compared [13]. The comparison provided the basis for developing the numerical setup and the experimental setup to check out subsonic flutter characteristics of some simple wing structures (e.g., a thin plate). Aeroelasticity on airplane wing, which had supercritical airfoil, was carried out in [18]. The model wings were made from different materials and dimensions. Hence, varied wing structures were created to accomplish a comprehensive analysis, and the flutter velocity was also restricted to appropriate values within the working range of experiment devices. At the same time, infinite element method with the help of the ANSYS software was also conducted to simulate the phenomena on the same model wings as a verification for the precision of the experimental models.

2. Two-way FSI method

2.1 FSI method

The generalized equations of aeroelasticity motion:

$$[M]\{\ddot{q}(t)\}+[C]\{\dot{q}(t)\}+[K]\{q(t)\}=\{F(t)\} \tag{1}$$

$$\{w(x,y,z,t)\}=\sum_{i=1}^{N} q_i(t)\{\phi_i(x,y,z,t)\} \tag{2}$$

where:
w: structural displacement at any time instant and position.
q: generalized displacement vector.
[M]: generalized mass matrice.
[C]: damping matrice.
[K]: stiffness matrices.
ϕ: normal modes of the structure.
N: total number of modes of the structure.
F: generalized force vector, which is responsible for linking the unsteady aerodynamics and inertial loads with the structural dynamics.

Eq. (1) shows that there are distinct terms representing the structure, aerodynamic, and dynamic disciplines. This equation was solved numerically by integrated CFD-CSD tool on the ANSYS software. This method was also called the two-way fluid-solid interaction (FSI). Aerodynamic loads were first calculated by CFD solver. Then, these loads were used to calculate the structural response of the wing structure through the fluid-solid interface. By using CSD solver, the structural deflection was estimated, and the mesh in each time step was deformed. The simulation of the two-way FSI is presented in **Figure 1** [10].

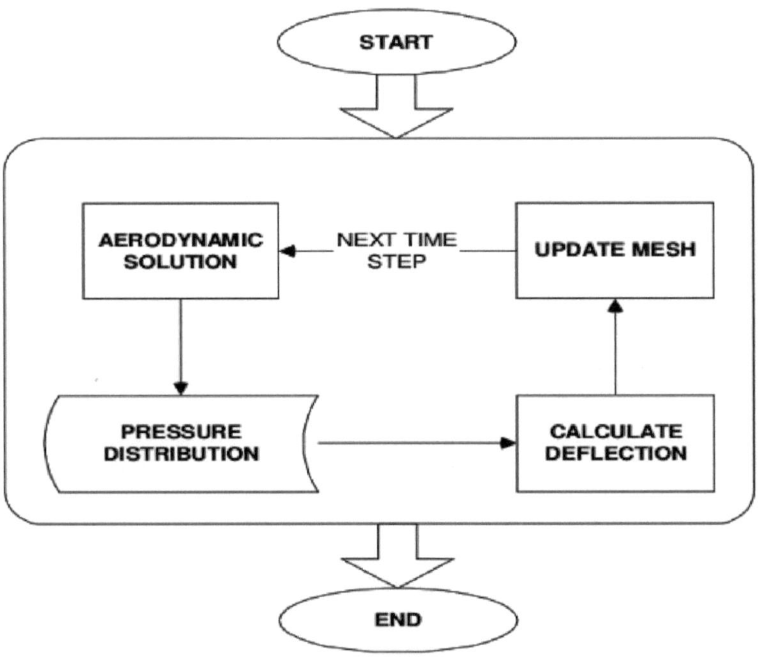

Figure 1.
Two-way FSI algorithm.

Following [8], the first four modes of vibration were sufficient to accurately model the wing structure response. So, in order to determine the time step of the unsteady problem, a modal method was applied to estimate first the natural frequency of the first four modes and then calculate the time step using the following formula:

$$\Delta t = \frac{1}{20f} \tag{3}$$

where:
Δt: time step
f: natural frequency

The flutter velocity was estimated from the vibration of the wing tip position, the most dangerous position of the wing [5, 8, 9, 13]. From the variation of this position, the damping coefficient except the influence of structural damping was measured. It means that there was an effect of aerodynamic damping coefficient on the vibration of the wing structure. The aerodynamic damping coefficient was calculated as follows:

$$\frac{1}{n}\ln\left(\frac{X_i}{X_{i+n}}\right) = \frac{2\pi\zeta}{\sqrt{1-\zeta^2}} \tag{4}$$

where:
X_i: i^{th} peak of vibration.
n: number of periods.
ζ: aerodynamic damping coefficient.

The damping coefficient increased/decreased when the air velocity decreased/increased. The type of vibration was distinguished from the value of damping coefficient:

- $\zeta > 0$: The vibration was convergent. The wing structure was stable.
- $\zeta < 0$: The vibration was divergent. The wing structure was unstable.
- $\zeta = 0$: The vibration was harmonic oscillation. The wing was in critical state. The air velocity was in the flutter velocity.

2.2 Wing model

AGARD 445.6 wing with aspect ratio of 1.65, taper ratio of 0.66, and 45° sweep angle at quarter chord line was studied as seen in **Figure 2**. The cross section of the wing was NACA65A004 airfoil in the stream-wise direction. This NACA65A004 airfoil was a symmetric airfoil with a maximum thickness of 4% of the local chord. The dimensions of the wing were root chord of 0.558 m, tip chord of 0.368 m, and semispan of 0.762 m. The wing model used in aeroelastic experiments [13] was constructed by laminated mahogany which was modeled as an orthotropic material with different material properties in different directions. The properties of the laminated mahogany are given in **Table 1**. The modal analysis was performed using mechanical APDLs solver to evaluate the accuracy of the constructed model.

The AGARD 445.6 wing was modeled at a zero attack angle and at altitudes of 9.65 and 14 km, the same conditions of experimental study in [13]. The wing was

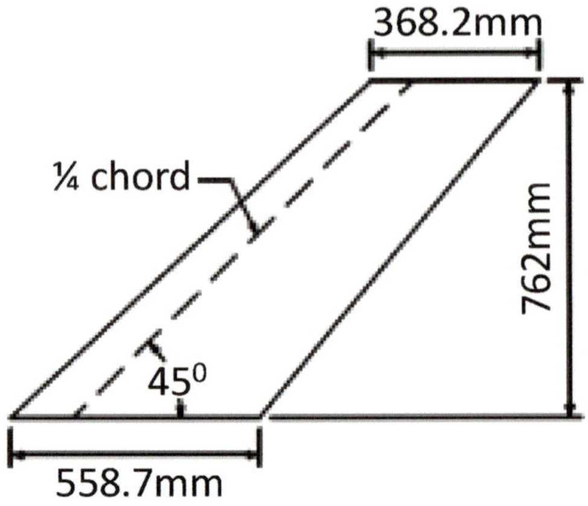

Figure 2.
Semispan AGARD 445.6 wing model.

Material property	E_{11}	E_{22}	E_{33}	G	υ	ρ
Value	3.151	0.416	0.416	0.439	0.310	397.5
Unit	GPa	GPa	GPa	GPa	N/m^2	kg/m^3

Table 1.
Mechanical properties for the weakened AGARD 445.6 wing.

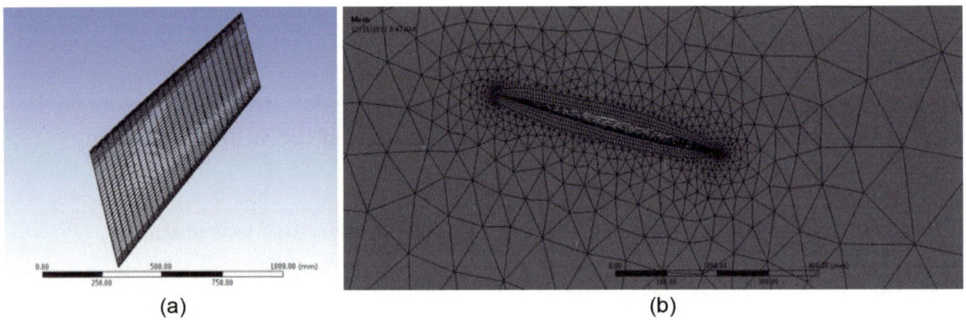

Figure 3.
Computational grids. (a) AGARD 445.6 wing grids. (b) Fluid computational grids.

meshed in 9257 nodes and 1350 elements (**Figure 3a**). The fluid domain in CFD problem was meshed in 67,949 nodes and 279,535 elements (**Figure 3b**).

The free-stream air velocity was from 0.29 to 0.59 M at an altitude of 9.65 km and from 0.47 M to 0.73 M at an altitude of 14 km.

2.3 Results

2.3.1 Modal analysis

The mode shapes are obtained from the finite element analysis of the modeled wing. The deflection contours between the modal analysis and experiment [13] were compared as shown in **Figure 4**. The natural frequencies between the developed solution, experiment [13], and other researches were also compared as shown in **Table 2**. It could be concluded that the obtained results were in good agreement

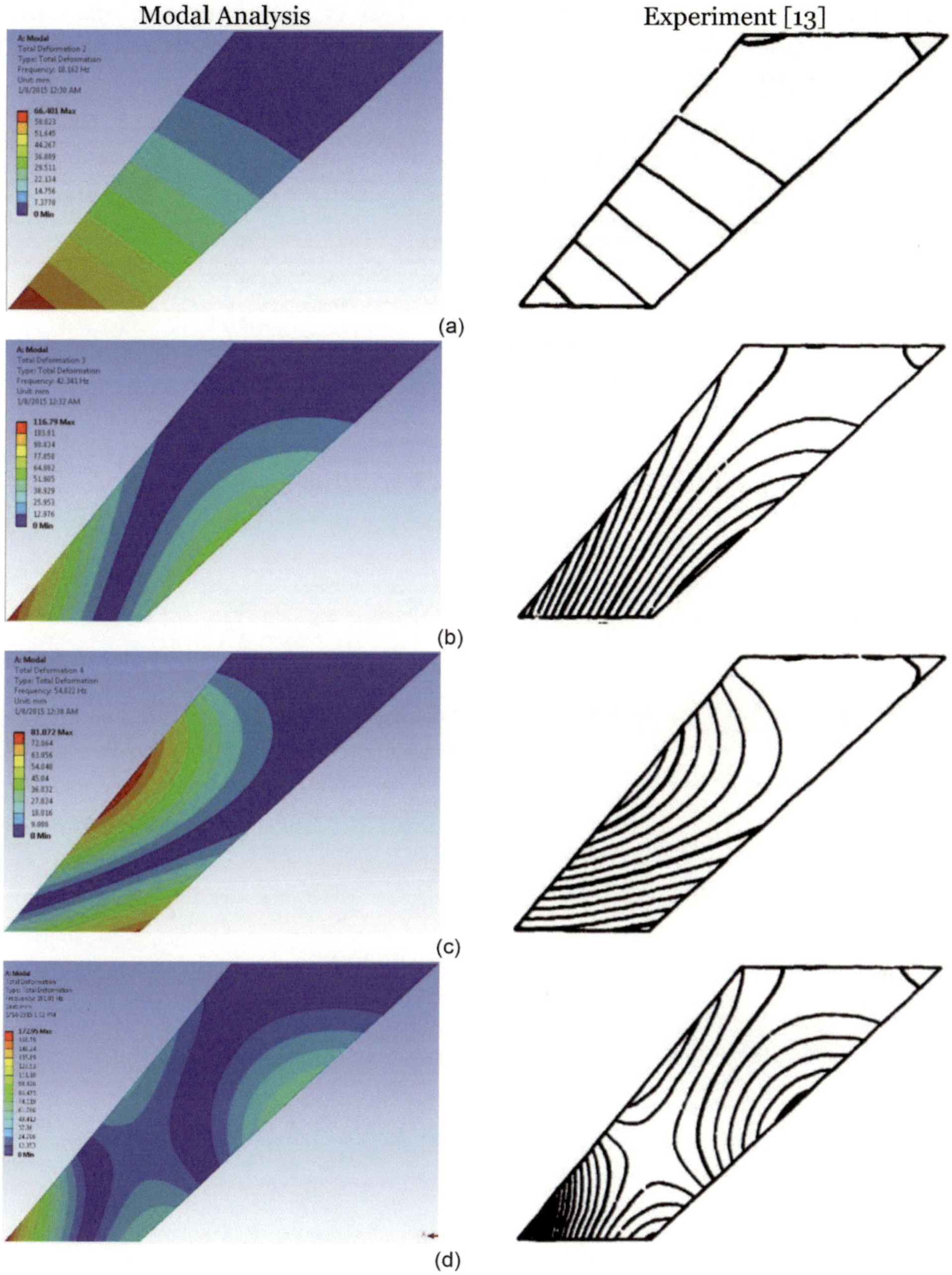

Figure 4.
Comparison of the mode shapes. (a) Mode 1. (b) Mode 2. (c) Mode 3. (d) Mode 4.

with the experimental results in [4, 6, 13] within an error relative of 8%. The frequency of the first mode was around 9.96 Hz. Following Eq. (3), the time step of the unsteady problem was estimated about 0.005 s.

2.3.2 Damping analysis

At altitude 9.65 km, the Mach number of air velocity was varied from 0.29 to 0.59 (M = 0.29; 0.35; 0.41; 0.47; 0.53; 0.59). From Eq. (4), the aerodynamic damping coefficient was calculated and presented in **Figure 5a**. The zero damping coefficients were interpolated at Mach number 0.46. While the experimental zero damping was 0.499 in [13], the difference between simulation results and results of [13] was about 8%.

	Mode 1	Mode 2	Mode 3	Mode 4
Modal analysis	9.96	41.5	52.95	99.99
Kolonay [4]	9.63	37.12	50.50	89.94
Erkut and Ali [6]	10.85	44.57	56.88	109.10
Yates [13]	9.6	38.1	50.7	98.5

Table 2.
Natural frequencies of modal analysis.

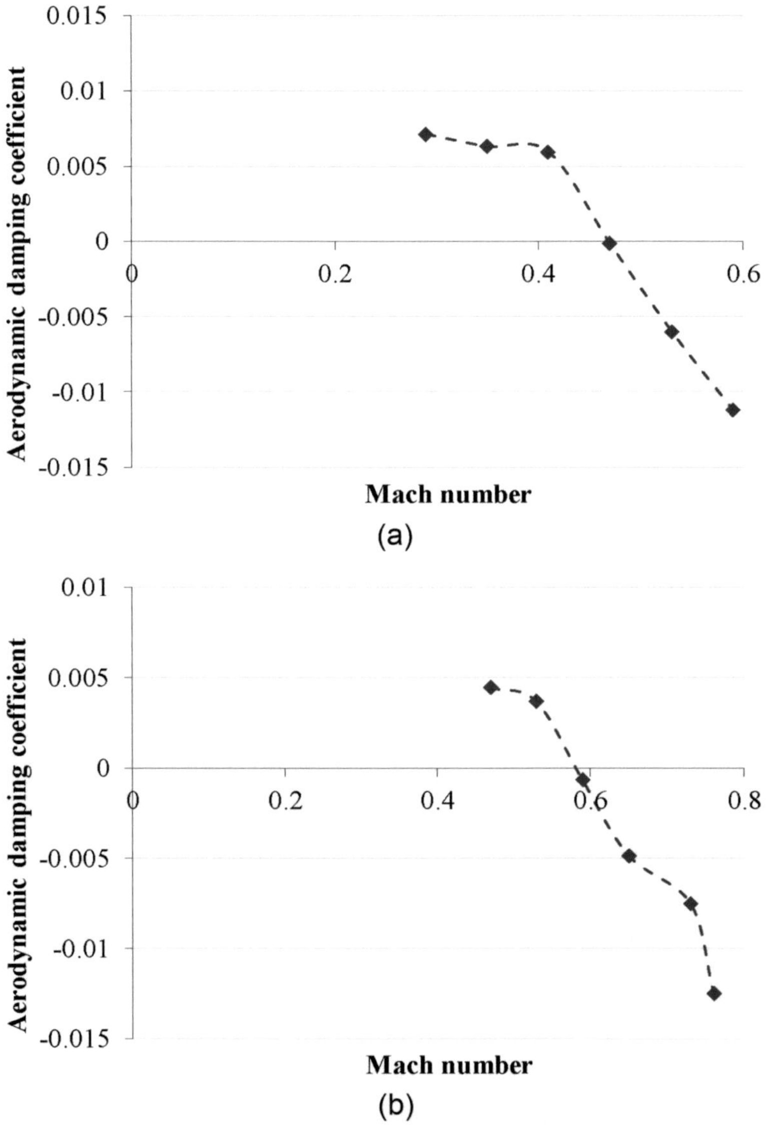

Figure 5.
Aerodynamic damping coefficients. (a) At altitude 9.65 km. (b) At altitude 14 km.

At altitude 14 km, the Mach numbers of air velocity were varied from 0.47 to 0.73 (M = 0.47; 0.53; 0.59; 0.65; 0.67; 0.73). From Eq. (4), the aerodynamic damping coefficient was calculated and presented in **Figure 5b**. The zero damping coefficients were interpolated at Mach number 0.58. While the experimental zero damping was 0.678 in [13], the difference within simulation results was about 14%.

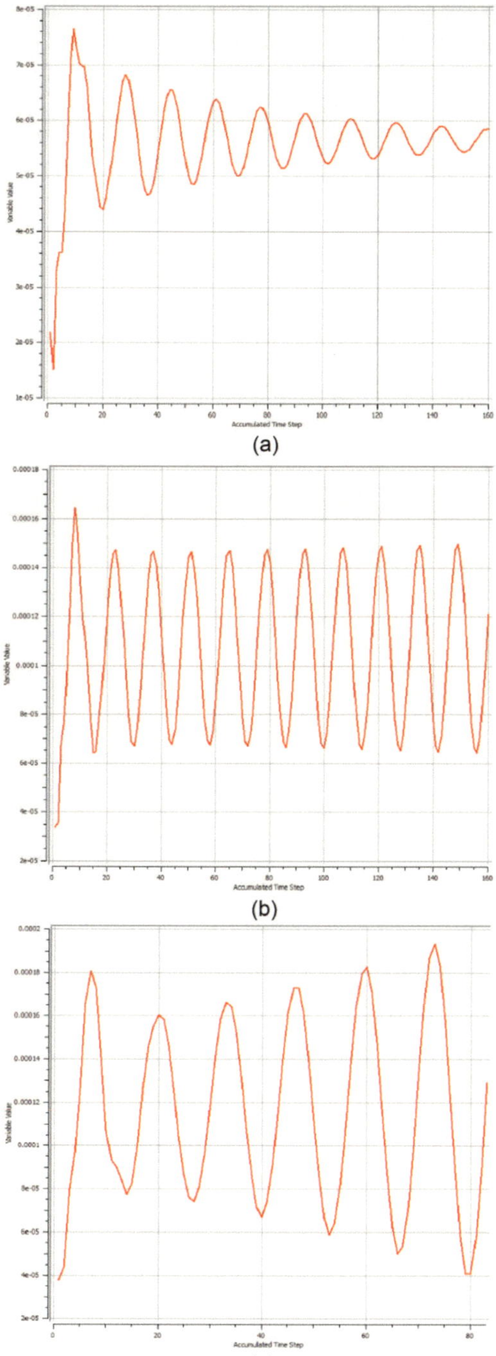

Figure 6.
Vibration of wing at altitude H = 9.65 km. (a) M = 0.29 – ζ = 0.0071. (b) M = 0.47 – ζ = −0.000125. (c) M = 0.53 – ζ = −0.006.

In both the two considered altitudes (9.65 and 14 km), the numerical results agreed well with the experimental results of [13] with a relative error about 14%. This difference would be from the computation such as the quality of mesh and order of model in CFD and CSD.

For more details of the stability of the wing structure, the vibrated value of the wing tip position at three Mach numbers were plotted as shown in **Figure 5**.

At a Mach number smaller than the flutter value, the damping coefficient was positive, and the wing was stable (**Figure 6a**).

At a Mach number near the flutter value, the damping coefficient was zero, and the vibration was harmonic oscillation (**Figure 6b**).

At a Mach number greater than the flutter value, the damping coefficient was negative, and the vibration was divergent (**Figure 6c**). This divergent vibration

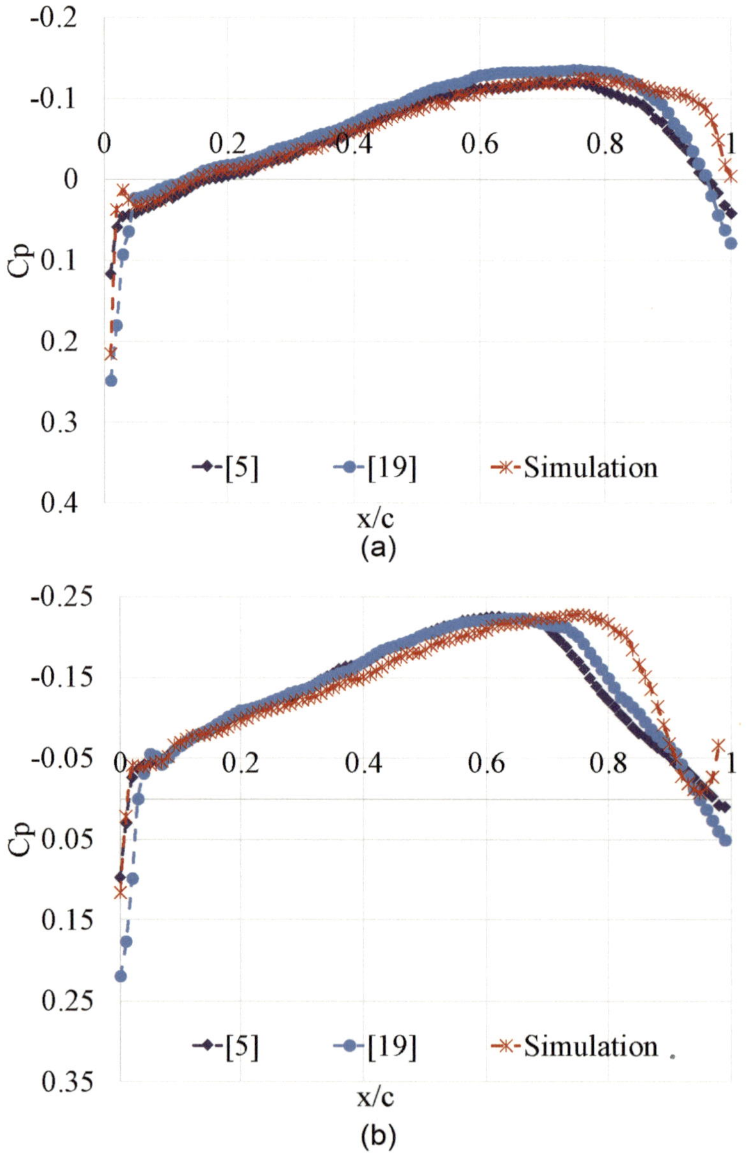

Figure 7.
Comparison of Cp distribution at M = 1.141, α = 0°. (a) 26 % semispan. (b) 75.5 % semispan.

would create the damage of the wing such as loss of control for the flap, aileron, fracture of wing, etc.

2.3.3 Simulation analysis

Dynamic aeroelastic analysis was a problem related to fluid-structure interaction over a period of time. Therefore, the quality of aerodynamic grid and the time step strongly influenced the results of aeroelastic analysis. These parameters were also two of the most important problems in the dynamic aeroelastic analysis.

In order to evaluate the quality of aerodynamic grid, the coefficient of pressure of AGARD 445.6 wing was first estimated at 26% semispan and at 75.5% semispan and then was compared with Ref. [6] as shown in **Figure** 7. The presented results were in good agreement with the results in Ref. [6]. It could be concluded that these simulation settings were appropriate for solving the transonic flow.

To evaluate the time step size, three different time sizes were examined such as 0.001, 0.002, and 0.005 s. As it could be seen in **Figure 8**, the displacement of the wing tip was reduced with the reduction of time step size up to 0.002 s until the aeroelastic simulation did not change [6]. Therefore, the value 0.002 s of time step size in the numerical solution was chosen for both aerodynamic and structural analysis.

The limit of flutter was identified by using damping estimations for a large test point at each Mach number. At M = 0.499, the oscillation of the displacement of the wing tip was harmonic (**Figure 9**), and it was considered as a flutter point. At this limit of flutter, the air speed was calculated as 174.26 m/s, and the density of air was calculated as 0.432 kg/m^3. These values were very close to the experimental values: 172.46 m/s for flutter speed and 0.428 kg/m^3 for density of air (**Table 3**).

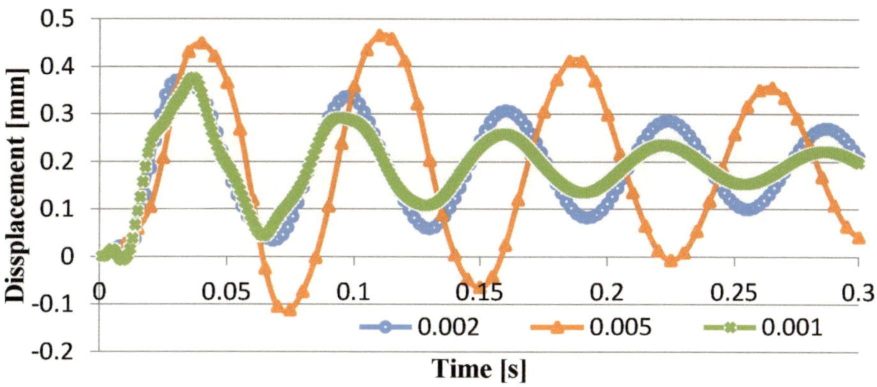

Figure 8.
Wing-tip oscillation depends on the selected time step.

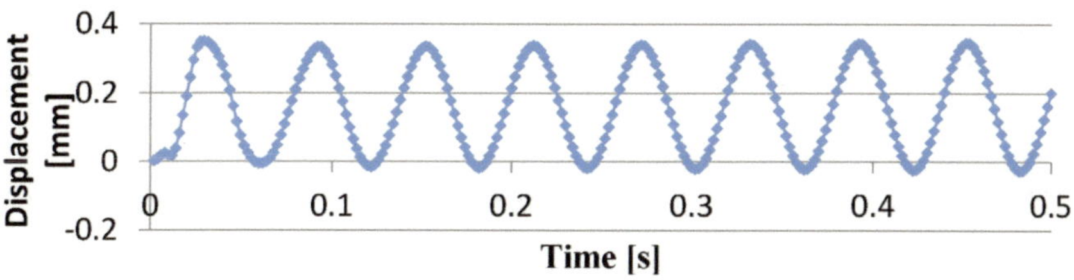

Figure 9.
Wing-tip oscillation of flutter point at M = 0.499.

	Velocity (m/s)	Density (kg/m^3)	Frequency at flutter (Hz)
Baskut [6]	171.84	0.3987	21.67
Yates [13]	172.5	0.42770	20.39
Simulation	174.257	0.43164	18.87
Relative error with [13]	1%	1%	7%

Table 3.
Flutter characteristics at M = 0.499.

This remark illustrated that the developed solution could be used to specify the transonic flutter characteristics with errors less than 10%.

3. Experimental method

3.1 Experimental setup

The test model was set in AF6116 (M = 0.1) subsonic wind tunnel located at the Hanoi University of Science and Technology, which was of a blowdown type with a closed test section ($0.4 \times 0.5 \times 1.0$ m^3). The wind speed could be arbitrarily varied up to 30 m/s, where the Reynolds number based on wing root chord was 10^6, which was driven by an 8 kW electric motor.

Flutter characteristics were determined with the help of the frequency meter and load cell, which allowed to specify the flutter frequency and root wing force, respectively. The oscillated frequency was measured by the DT-2234C+ frequency meter. The signal of measured frequency was averaged by five measurements. The force applied to the wing was measured by load cell system. In the experimental aeroelastic analysis, the flutter frequency and the flutter amplitude were measured at different velocities ranging from 10 to 30 m/s using an oscillator generator system.

3.2 Wing model

Two wing models with the parameter and dimension are shown in **Figure 10** and **Table 4**. The non-structure wing had only balsa wood, while the structure wing had balsa wood for skin and carbon rod and hard wood for the inner parts.

3.3 Results

Experimental results showed that a flutter phenomenon appeared with the non-structure wing (broken wing in **Figure 10c**), but this phenomenon did not happen with the structure wing model. It could be explained by the more durability of structure wing than that of the non-structure wing with the testing range of velocity. Experiments also demonstrated that the combination of multiple materials to more durability of structure of wing could be highly effective in preventing flutter phenomenon [18].

The measurement results of the non-structure wing were shown in **Table 5**. When the attack angle increased, the velocity of first oscillation, flutter velocity, and frequency decreased.

Figure 11 resumed the measurement of the force at the wing root in varying velocities from zero to flutter velocity and more by using the load cell system. After increasing the air velocity from zero to the limit of non-structure wing, the limit

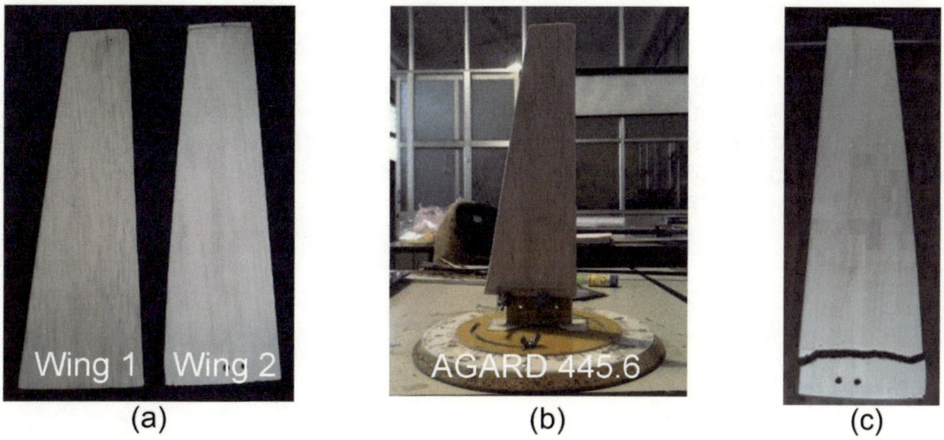

Figure 10.
Experimental models. (a) Wing model. (b) Wing with support. (c). Broken wing.

	Wing 1 (non-structure)	Wing 2 (with structure)
Chord length	300 mm	300 mm
Root chord length	500 mm	500 mm
Tip chord length	100 mm	100 mm
Profile	NACA65A004	NACA65A004
Taper ratio	0.5	0.5
Material	Balsa wood	Balsa wood, carbon rod, hard wood
Mass	5.1 g	11.6 g

Table 4.
Wing models.

Attack angle (^{0})	Velocity of first oscillation (m/s)	Flutter velocity (m/s)	Frequency (Hz)
0	20	22	37.38
5	17.8	20.5	32.22
10	15.2	19.5	30.31

Table 5.
Flutter characteristics at different attack angles—Non-structure wing.

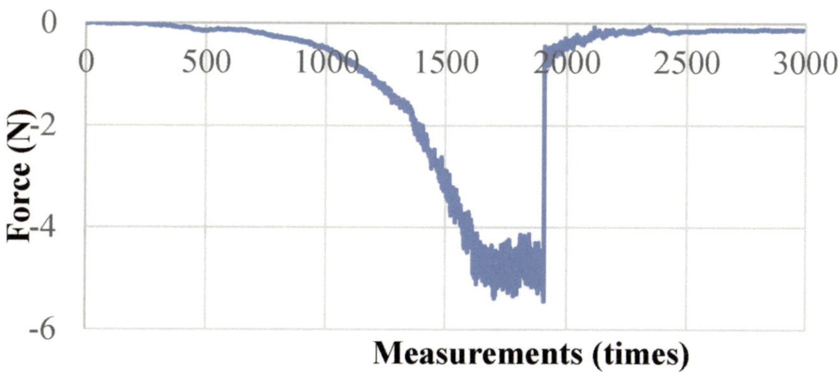

Figure 11.
Force at the wing root of the non-structure wing—Attack angle 10°.

velocity of non-structure wing was found at 19.5 m/s. The load at the wing root of this wing at flutter velocity is shown in **Figure 12**. The maximum force was 5.44 N, and the minimum force was 4.32 N.

With structure wing, different modes of vibration appeared depending on the characteristics of the structure as remarked in [8]. With the help of the oscillator generator system, the specific oscillation frequencies of the first four modes were estimated as shown in **Table 6**.

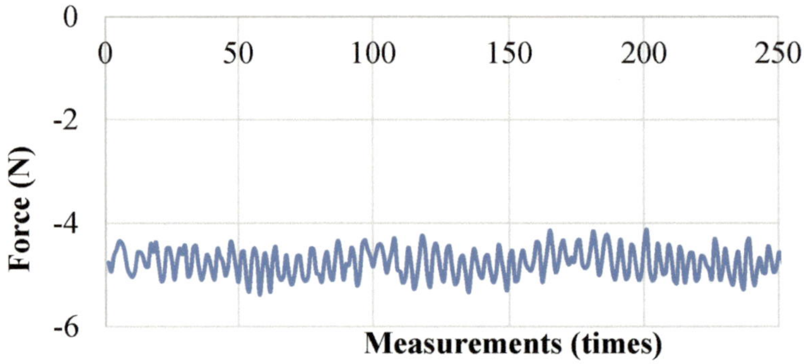

Figure 12.
Force at the wing root of the non-structure wing—Attack angle 10° and velocity 19.5 m/s.

Mode	Frequency (Hz)	
	Non-structure wing	Structure wing
1	24.3	23.5
2	67.0	63.0
3	103.6	115.3
4	132.0	139.0

Table 6.
Specific oscillation frequency of the non-structure and structure wings.

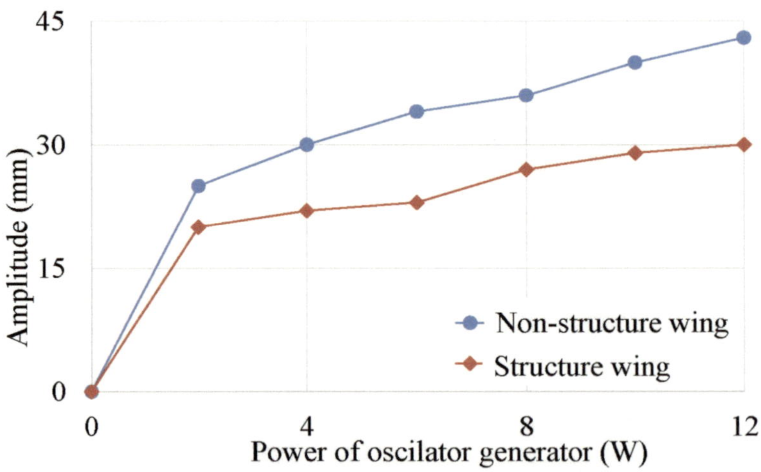

Figure 13.
Amplitude of oscillation of the structure wing—Mode 1.

The frequencies of the first and second modes of the non-structure wing were higher than those of the structure wing, while the frequencies of third and fourth mode of structure wing were higher than those of non-structure wing (**Table 6**). Considering both wings in the first mode of oscillation, when the force was applied to the wing, the amplitude of the non-structure wing was higher than that of the structure wing (**Figure 13**). In conclusion, the non-structure wing was easier to resonate than the structure wing.

4. IBM method

4.1 Methodology

Fluid flow and deformation of structure were governed by the following equations with assumption of linear elastic structure [11]:

$$\frac{\partial u}{\partial t}=\frac{1}{\text{Re}}\nabla^2 u-u\nabla u-\nabla p+f \tag{5}$$

$$\nabla u=0 \tag{6}$$

$$\frac{d\left(m_p U_c\right)}{dt}=F+g \tag{7}$$

$$\frac{dx_c}{dt}=U_c \tag{8}$$

$$\frac{d\left(I_p\cdot\omega_p\right)}{dt}=T \tag{9}$$

$$\frac{d\theta_p}{dt}=\omega_p \tag{10}$$

where:
u: fluid velocity vector
p: fluid pressure
f: force that affected on wing
Re: Reynolds number
U_c: displacement velocity
ω_p: angular velocity
x_c: center of gravity
θ_p: rotation of wing r
mp: mass of wing
Ip: inertial moment of wing
F: force created by fluid passing through the wing
T: moment created by fluid passing through the wing

To solve out these equations using IBM method, the most important was that the velocity of fluid at fluid-solid interface was equal to the velocity of the wing. It means that the interaction force (f) between the wing and fluid was calculated such that the boundary condition of fluid was satisfied on the surface of the wing.

IBM method used the Cartesian grid and immersed boundary that were illustrated in **Figure 14**, in which the moving surface of wing was described by

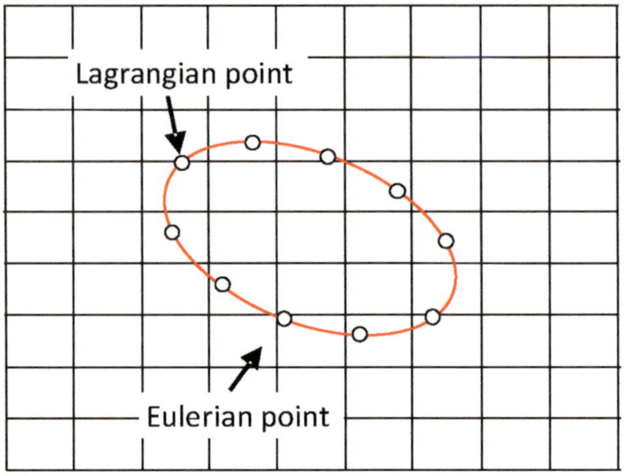

Figure 14.
Cartesian grid and immersed boundary.

Lagrangian points (rounded points) and fixed points in fluid were called Eulerian points. Parameters of Lagrangian points were noted as capitalization.

Discrete partial derivative of velocity over time in Eq. (5) with denote intermediate velocity at zero force of Lagrangian point, $\hat{U}_k$, force F of Lagrangian points were estimated as follows:

$$F = \frac{U^{k+1} - \hat{U}^k}{\Delta t} - \left(\frac{1}{\text{Re}} \nabla^2 U - U \nabla U - \nabla P \right) \quad (11)$$

where:

k: time step.

U_{k+1}: identified from the moving surface of wing, so this velocity was known as $U^{(b)}$.

Force was created from displacement and affected on the fluid element. Force was calculated using the following interpolation formula:

$$f = \sum_{l=1}^{N} F(x_l)\,\delta_h(x - x_l)\,\Delta V_l \quad (12)$$

where:

x: coordinate of Eulerian point.

N: set of Lagrangian points round Eulerian point *l*.

x_l: coordinate of Lagrangian point.

ΔU_l: volume of effect corresponding to Lagrangian point *l*.

δ_h: 3D delta function identified as follows:

$$\delta_h(x - x_l) = \delta_h^{1D}(x - x_l)\,\delta_h^{1D}(y - y_l)\,\delta_h^{1D}(z - z_l) \quad (13)$$

$$\delta_h^{1D}(r) = \frac{1}{h}\varphi(r) \quad r = \frac{x - x_l}{h} \quad (14)$$

$$\varphi(r) = \begin{cases} \frac{1}{6}\left(5 - 3|r| - \sqrt{1 - 3(1 - |r|)^2}\right) & \frac{1}{2} \leq |r| \leq \frac{3}{2} \\ \frac{1}{3}\sqrt{1 - 3|r|^2} & |r| \leq \frac{1}{2} \\ 0 & |r| \geq \frac{3}{2} \end{cases} \tag{15}$$

Three-step Runge–Kutta method was applied to solve Navier-Stokes equations and Newton equations as follows:

- Step 1: Calculate the instantaneous velocity at Eulerian points with no immersed boundary surface, i.e., f = 0:

$$\tilde{u} = u^{k+1} + \Delta t\left(2\alpha_k \upsilon \nabla^2 u^{k-1} - 2\alpha_k \nabla p^{k-1} - \gamma_k\left((u\cdot\nabla)u\right)^{k-1} - \zeta_k\left((u\cdot\nabla)u\right)^{k-2}\right) \tag{16}$$

where:
k: step calculation of Runge-Kutta method (k = 1, 2, 3)
α_k: coefficient of kth step calculation
γ_k: coefficient of kth step calculation
ζ_k: coefficient of kth step calculation
υ: kinematic viscosity
Apply this instantaneous velocity to calculate Lagrangian velocity on the surface of the wing:

$$\tilde{U}_{x_l} = \sum_{k=1}^{N} \tilde{u}_x \delta_h\left(x - x_k\right) h^3 \tag{17}$$

This Lagrangian velocity was combined with wing velocity, $U^{(b)}{}_{xl}$, which was calculated from the dynamic equation of wing, to calculate forces of Lagrangian points (F) following Eq. (11). Then, apply this force of Lagrangian points to Eq. (12) to calculate force of Eulerian points (f).

- Step 2: Solve out Navier-Stokes Eqs. (5) using calculated forces of Eulerian points to estimate the effect of flutter of the wing into the velocity field around the wing:

$$\nabla^2 u^* - \frac{u^*}{\alpha_k \upsilon \Delta t} = -\frac{1}{\upsilon \alpha_k}\left(\frac{\tilde{u}}{\Delta t} + f_x\right) + \nabla^2 u^{k-1} \tag{18}$$

To satisfy the continuity equation, a temporary pressure was described:

$$\nabla^2 \Phi^k = \frac{\nabla \cdot u^*}{2\alpha_k \Delta t} \tag{19}$$

- Step 3: Solve Eq. (17), and calculate velocity and pressure at kth step of the Runge-Kutta method:

$$u^k = u^* - 2\alpha_k \Delta t \nabla \Phi^k \tag{20}$$

$$p^k = p^{k-1} + \Phi^k - \alpha_k \Delta t \nabla^2 \Phi^k \quad (21)$$

From the estimated forces at Lagrangian points, translational and angular movements of the wing were carried out by solving Eqs. (7) and (9):

$$U_c^k = U_c^{k-1} + 2\alpha_k \frac{\Delta t}{M}(F + G) \quad (22)$$

$$\omega_p^k = \omega_p^{k-1} + 2\alpha_k \frac{\Delta t}{I_c} T \quad (23)$$

After calculating the velocity of center of gravity (U_c^k) and angular velocity of wing (ω_p^k), coordinates of Lagrangian points were estimated by the same expressions.

4.2 Wing model

Four different wing models were carried out in order to analyze the effect of the wing structure. There were two rectangular and two trapezoid 3D-shape wings, and each 3D-shape wing had symmetric (NACA65A004) and asymmetric (supercritical) airfoil, respectively. The wings had the same area of 450 cm^2 and the same semispan-wise length of 30 cm. Therefore, the rectangular wing had a chord length of 7.5 cm, while the trapezoidal wing has no sweep angle, and the leading edge line had a tip chord length of 5 cm and root chord length of 10 cm. The wings were made of aluminum (**Figure 15**).

Experiments were performed using a low-speed blowdown wind tunnel, which belongs to the Department of Aerospace Engineering at the Hanoi University of Science and Technology, Vietnam. This wind tunnel had a maximum free-stream velocity in empty test section of 30 m/s that corresponded to Reynolds number 10^6. The wind tunnel was operated continuously by an 8 kW electric motor. The turbulence level in test section was slightly less than 1%. Free-stream velocity was kept constant in test section within ±2%. Total pressure of free-stream and dynamic pressures were measured by pitot tube within ±2%. Ambient temperature was measured within ±1%. Both experimental and numerical researches were performed at air velocity of 20 m/s and attack angle of 5°.

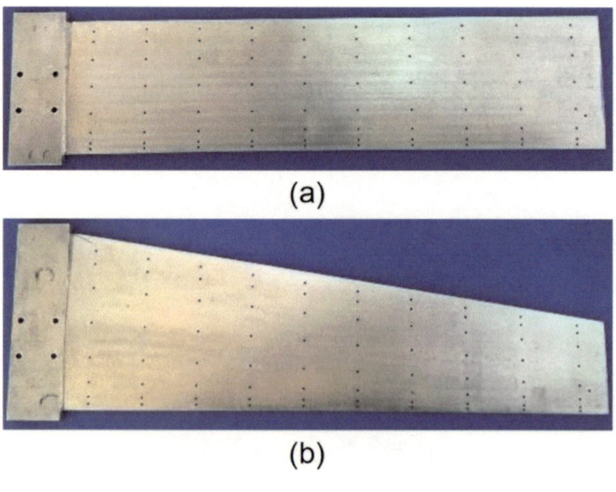

Figure 15.
Wing models. (a) Rectangular wing. (b) Trapezoidal wing.

For the experimental study, 160 pressure taps were applied on the wing model (**Figure 15**). These pressure taps were connected to an external digital manometer via stainless and silicon tubes. Each pressure tap was measured one time with

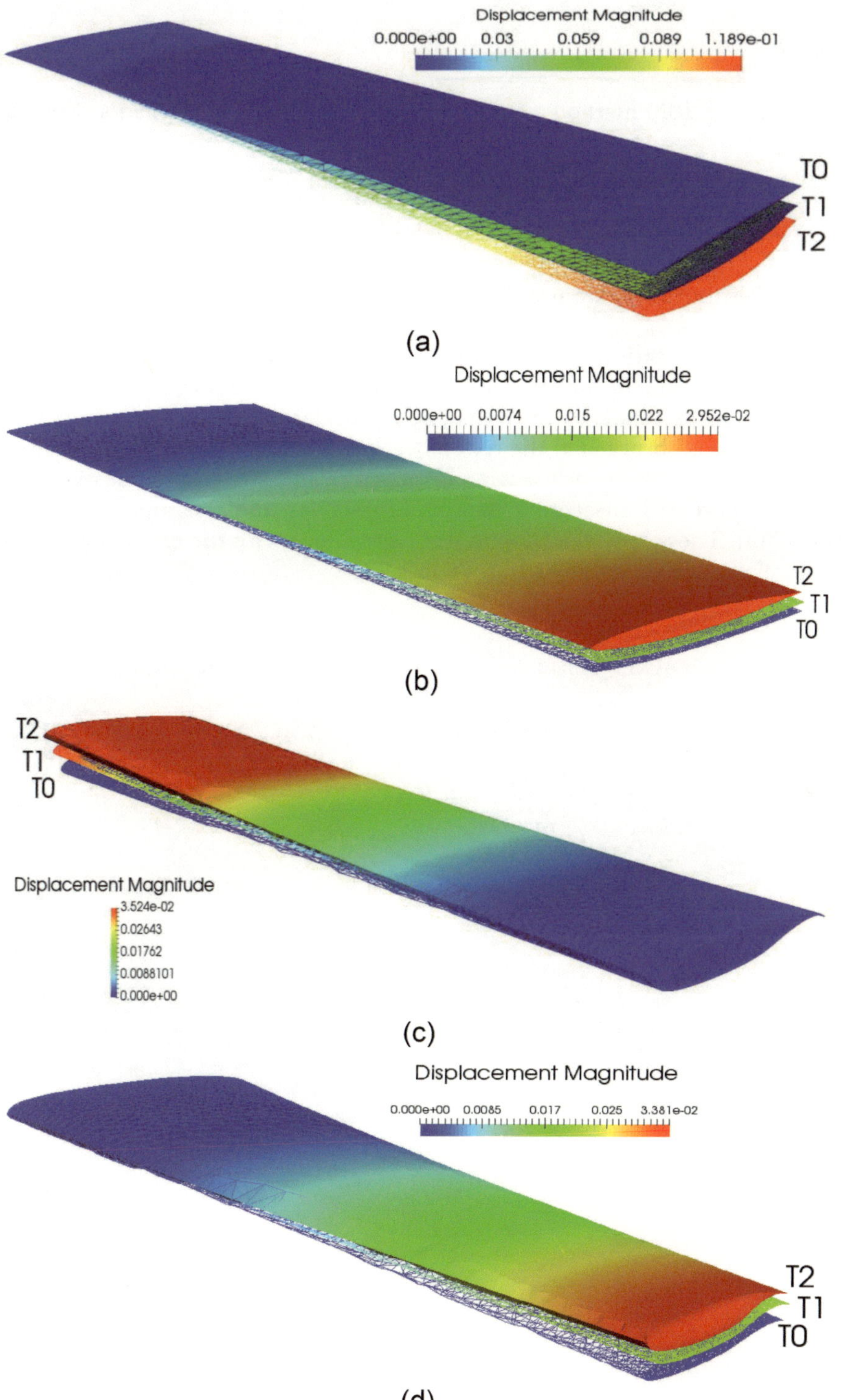

Figure 16.
Instant displacement of wings. (a) NACA65A004 – Rectangular wing. (b) NACA65A004 – Trapezoidal wing. (c) Supercritical – Rectangular wing. (d) Supercritical – Trapezoidal wing.

waiting time of 5 s (average of about 1000 instant values) using the Keyence pressure measurement. The standard deviation of the Keyence pressure measurement errors was within ±0.001 Pa. Moreover, flutter of wing was captured with help of high performance camera.

4.3 Results

The results of IBM method were analyzed at three different instants (**Figure 16**):

- Time T_0: initial time when distortion did not occurred
- Time T_1: time between maximum deformation and non-deformation
- Time T_2: time of maximum distortion

The wing deformation was maximum at the tip of the wing and decreased gradually into the root of the wing over time. However, the normal stress was found to have an opposite tendency in comparison with deformation. The maximum normal stress was observed at the root of the wing, while the minimum normal stress was found at the tip of the wing (**Figures 17** and **18**). It could be explained by the fixed support with fuselage at the root of the wing and free support at the tip of the wing [13]. These remarks were in well agreement with the experimental results within a relative error less than 10% (**Table 7**).

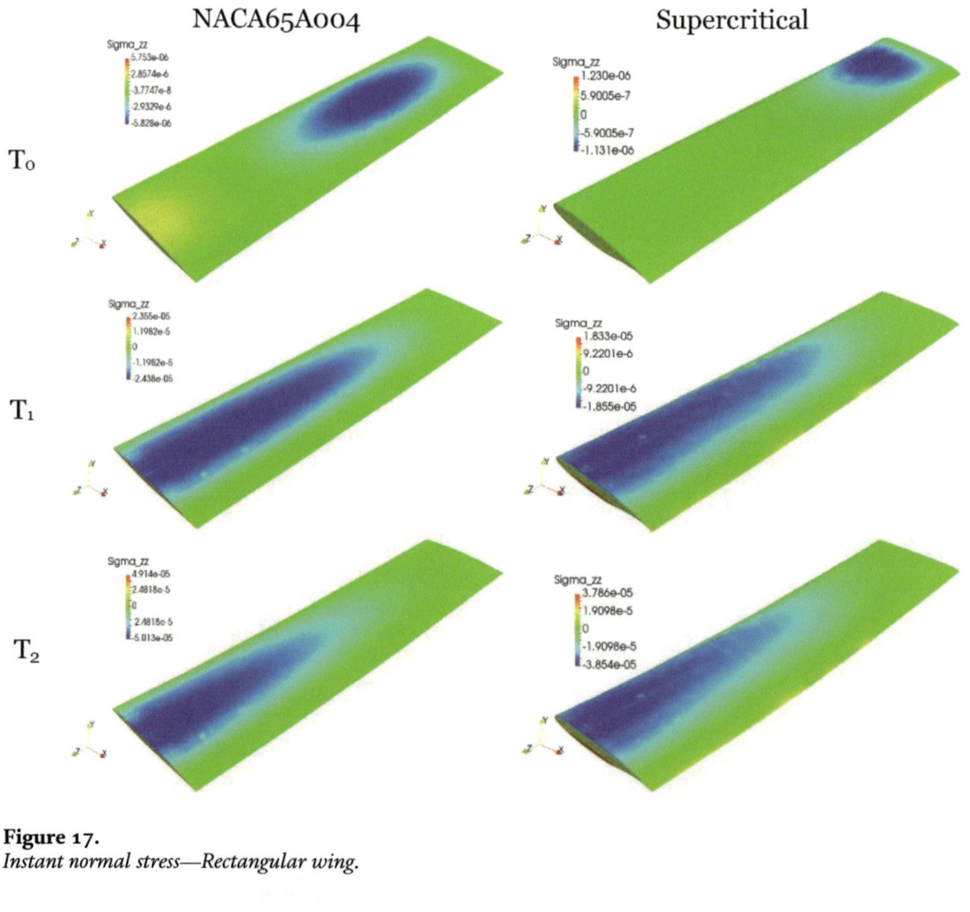

Figure 17.
Instant normal stress—Rectangular wing.

At T0 instant, the normal stress had important value near the wing tip. During flutter behaviors of wing, this important normal stress propagated from the tip of the wing to the root of the wing. The maximum value of the normal stress was found out at the root of the wing and at T_2 instant.

With the same airfoil, the rectangular wing was found to be more distorted and have higher maximum deformation and higher maximum normal stress than the trapezoid wing. Thus, 3D-shape wing contributed significantly to the deformation of wing when aeroelasticity phenomenon occurred (**Table 7**).

With the same 3D-shape wing, the maximum deformation and maximum normal stress of NACA65A004 rectangular wing were higher than those of the rectangle supercritical wing. Meanwhile, the maximum deformation and maximum normal stress of NACA65A004 trapezoidal wing were less than those of the supercritical trapezoidal wing. It could be concluded that the 3D shape of wing played an important role in the durability of the structure (**Table 7**).

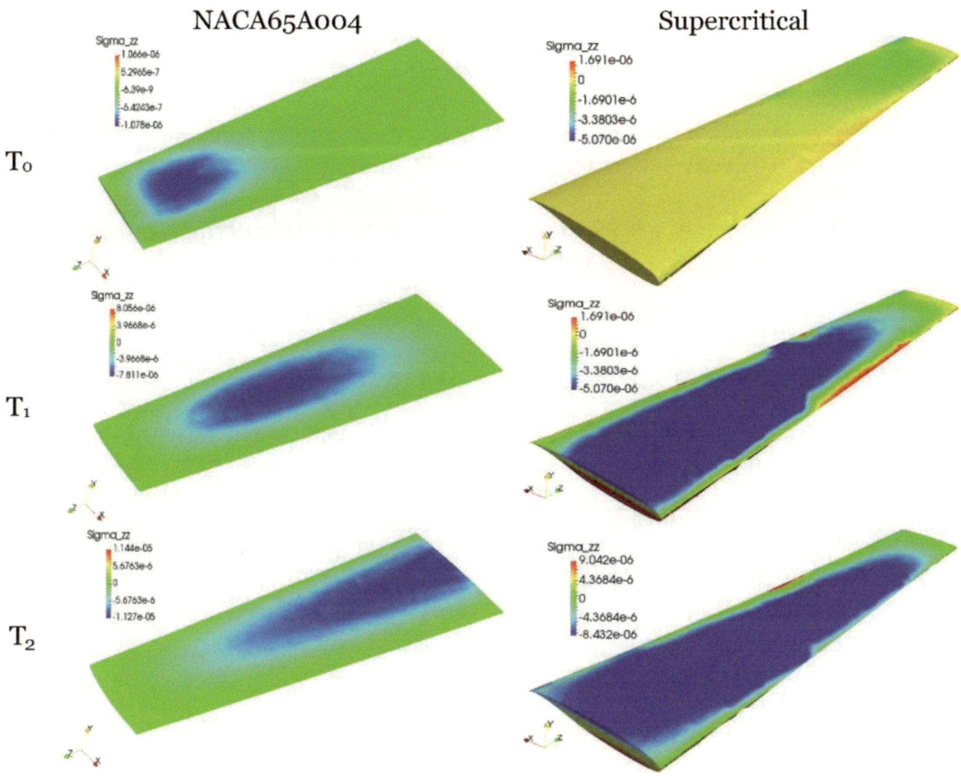

Figure 18.
Instant normal stress—Trapezoidal wing.

Wing	IBM method (mm)	Experiment method (mm)	Relative error (%)
NACA65A004-rectangular	0.119	0.131	9.7
NACA65A004-trapezoidal	0.030	0.033	8.6
Supercritical-rectangular	0.035	0.039	9.9
Supercritical-trapezoidal	0.034	0.037	9.2

Table 7.
Maximum deformation of the wing tip.

5. Conclusions

The flutter phenomenon of AGARD 445.6 wing was determined by (a) a modal approach for a structural response; (b) an aerodynamic damping coefficient to predict the appearance of flutter phenomenon; (c) a strongly coupled FSI method to predict the aeroelastic response for subsonic and transonic flutter characteristics; (d) an experiment method to predict the aeroelastic response for subsonic flutter characteristics with wing structure; (e) and an IBM method to improve the interface between the fluid and solid of aircraft wing.

In brief, the major results could be summarized as follows:

- Experimental results were in good agreement with numerical results within a relative error less than 10%.

- During aeroelasticity phenomenon, deformation of the wing tip was maximum while it was minimum at the wing root. The tendency of normal stress was in contrast with deformation. The minimum normal stress was observed at the wing tip, while the maximum normal stress was observed at the wing root;

- Geometry of wing (3D shape, airfoil) had a significantly contribution to the deformation of wing when aeroelasticity phenomenon occurred.

For further research of aeroelasticity in the future, both experimental and numerical researches at low and high speed should be performed.

Acknowledgements

A part of this work was supported by the Ministry of Science and Technology in Vietnam through the bilateral and multilateral research project HNQT/SPĐP/12.19.

Appendices and nomenclature

w	structural displacement at any time instant and position
q	generalized displacement vector
[M]	generalized mass matrice
[C]	damping matrice
[K]	stiffness matrices
ϕ	normal modes of the structure
N	total number of modes of the structure
F	generalized force vector
X_i	i^{th} peak of vibration
n	number of period
ζ	aerodynamic damping coefficient
M	Mach number
u	fluid velocity vector
p	fluid pressure
f	force that affected on wing
Re	Reynolds number
U_c	displacement velocity
ω_p	angular velocity

x_c	center of gravity
θ_p	rotation of wing
m_p	mass of wing
I_p	inertial moment of wing
F	force that created by fluid go pass through the wing
T	moment that created by fluid go pass through the wing
X	coordinate of Eulerian point
N	set of Lagrangian points round Eulerian point l
x_l	coordinate of Lagrangian point
ΔU_l	volume of effect corresponded to Lagrangian point l
δ_h	3D delta function
$\alpha\ \gamma\ \zeta$	coefficient of step Runge-Kutta calculation
υ	kinematic viscosity
E_{11}	normal stress following x-coordinates
E_{22}	normal stress following y-coordinates
E_{33}	normal stress following z-coordinates
G	shear stress
ρ	density
Cp	coefficient of pressure
α	attack angle

Author details

Hoang Thi Kim Dung[1]* and Nguyen Phu Khanh[2]

1 School of Transportation Engineering, Hanoi University of Science and Technology, Hanoi, Vietnam

2 Faculty of Vehicle and Energy Engineering, Phenikaa University, Hanoi, Vietnam

*Address all correspondence to: dung.hoangthikim@hust.edu.vn

References

[1] Edwards JW. Computational Aeroelasticity Challenges and Resources. NASA Technical Report. N89-19264; 1989

[2] Liu DD, Sarhaddi D, Piolenc FM. Flutter Prevention Handbook: A Preliminary Collection. Technical report WL–TR–96–3111, Wright Laboratory; 1996

[3] Ramsey JK. NASA Aeroelasticity Handbook Volume 2: Design Guides Part 2. Technical Paper NASA/TP—2006-212490/VOL2/PART2, NASA Glenn Research Center; 2006

[4] Kolonay RM. Unsteady aeroelastic optimization in the transonic regime [Thesis]. Purdue University; 1996

[5] Lee-Rausch EM, Batina JT. Calculation of AGARD Wing 445.6 Flutter using Navier-Stokes Aerodynamics. AIAA Paper No. 93–3476; 1993

[6] Başkut E, Akgül A. Development of a closely coupled procedure for dynamic aeroelastic analyses. Scientific Technical Review. 2012;**62**(2):30-39

[7] Pahlavanloo P. Dynamic Aeroelastic Simulation of the AGARD 445.6 Wing using Edge. Technical report FOI-R–2259—SE; 2007

[8] Chen X, Zha G-C, Yang M-T. Numerical simulation of 3-D wing flutter with fully coupled fluid-structural interaction. AIAA Paper 2006–0697; 2006

[9] Liu F, Cai J, Zhu Y, Tsai HM, Wong ASF. Calculation of wing flutter by a coupled fluid-structure method. Journal of Aircraft. 2001;**38**(2):334-342

[10] Hoang TKD, Nguyen PK, Tran NM. Study of wing flutter by a coupled fluid-structure method. In: Proceeding of 14th Asia Congress of Fluid Mechanics. Vol. 1. 2013. pp. 144-148

[11] Hoang TKD, Pham VS, Tran NK, Nguyen DC, Nguyen PK. Aeroelastic analysis on wing structure using immersed boundary method. In: Lecture Notes in Mechanical Engineering. Springer; 2018. pp. P783-P792. DOI: 10.1007/978-981-10-7149-2_55

[12] Dugundji J, Dowell E, Perkin B. Subsonic Flutter of Panels on Continuous Elastic Foundations - Experiment and Theory. ASRL Technical Report No.74 4; 1962

[13] Yates EC. AGARD standard aeroelastic configurations for dynamic response I-wing 445.6. AGARD Report 765; 1988

[14] Ricketts RH. Experimental Aeroelasticity - History, Status and Future in Brief. NASA Technical Memorandum 102651; 1990

[15] Heinze S. Aeroelastic Concepts for Flexible Aircraft Structures [Thesis]. Royal Institute of Technology; 2007

[16] Achuthan CP, Shanthini G, Shivaprasad MV, Chandra N, Manjuprasad M. Dynamic response and flutter prediction of structures with unknown system parameters using experimental modal parameters. Innovative Systems Design and Engineering. 2011;**2**(4):23-39

[17] Nguyen PK, Mori K, Hoang TKD, Nguyen VH, Nguyen HA. Research on simulation and experiment of dynamic Aeroelastic analysis on wing structure. Applied Mechanics and Materials. 2015; **798**:541-545. DOI: 10.4028/www/scientific.net/AMM.798.541

[18] Tran NK, Dang DC, Dao DH, Nguyen PK, Hoang TKD. Experimental research on the effect of wing structure on aeroelasticity phenomenon. Applied Mechanics and Materials. 2019;**889**: 403-409. DOI: 10.4028/www.scientific.net/AMM.889.403

Section 2

Optimization

Chapter 4

Optimization of Lift-Curve Slope for Wing-Fuselage Combination

Vladimir Frolov

Abstract

The paper presents results obtained by the author for wing-body interference. The lift-curve slopes of the wing-body combinations are considered. A 2D potential model for cross-flow around the fuselage and a discrete vortex method (DVM) are used. Flat wings of various forms and the circular and elliptical cross sections of the fuselage are considered. It was found that the value of the lift-curve slopes of the wing-body combinations may exceed the same value for an isolated wing. An experimental and theoretical data obtained by other authors earlier confirm this result. Investigations to optimize the wing-body combination were carried within the framework of the proposed model. It was revealed that the maximums of the lift-curve slopes for the optimal midwing configuration with elliptical cross-section body had a sufficiently large relative width (more than 30% of the span wing). The advantage of the wing-fuselage combination with a circular cross section over an isolated wing for wing aspect ratio greater than 6 can reach 7.5% at the relative diameter of fuselage equal to approximately 0.2.

Keywords: wing-fuselage combination, lift-curve slope, discrete vortex method, 2D potential cross-flow model, optimization

1. Introduction

An analysis of a lift-curve slope for wing-fuselage combinations currently plays an important role in studies of aerodynamics and the preliminary design of a modern aircraft.

Since the aircraft occurrence aircraft designers have been interested in the problems of the wing-body interference in aviation and missile technology. Initially, research is focused on the experimental study of specific wing-body combinations [1–6]. First mathematical models of the wing-fuselage interference were offered later. The solution of the linearized problem of the ideal incompressible flow around arbitrary shape wings in the presence of the fuselage is a difficult task since it is necessary to solve the three-dimensional Laplace equation for the velocity potential which satisfies the boundary conditions on the surface of the wing-body combination and the boundary conditions at infinity. One of the few exact solutions was obtained by Golubinsky in the article [7]. The first theoretical calculations were based on the inversion of discrete vortices inside the cross-section body [8], on the solution of integral equations [9], on the application the thin body theory [10–21] or the strip method [14, 15, 22], and on the application of the velocity potential [23–26] or the stream function [27] written in the Trefftz's plane. The application of the velocity potential in the Trefftz's plane [22–26, 28–30] gives the opportunity to get the distribution of lift along the wingspan. Let us pay attention to one important result that was first theoretically obtained by Multhopp and presented in a review by Ferrari [22]. It is important to note that this fact was experimentally confirmed by Jacobs and Ward [1]. This result shows that the value of the lift-curve slope of the wing-body combination at a certain relative value diameter of the fuselage $\overline{D} = d_f/b$ (b is the wingspan; d_f is the diameter of the fuselage) exceeds the same value for the isolated wing of the same geometry which is used in the wing-fuselage configuration. Let us give some examples confirming this fact. For the scheme of midwing monoplane with cylindrical fuselage $\overline{D} = 0.14$ and trapezoidal wing with aspect ratio $AR = 4.83$ and taper ratio $\lambda = 2, 38$, the value of the lift-curve slope of the wing-body combination $C_{L_a\mathrm{W,\,B}}$ exceeds the same value $\overline{C}_{L_aW}$ for the isolated wing of the same geometry which is used in the wing-body configuration by approximately 5% (4.75%, an experiment; 4.92%, theory) [22]. For the wing-body combination No. 13 in the experimental Jacobs and Ward's paper [1], the relative increase in the value of the lift-curve slope of the wing-fuselage combination was slightly greater (≈6.49%) than the isolated wing of the same geometry, which was obtained also. In Korner's book [6], it was noted that the value of the lift-curve slope of the wing-body combination is approximately 5% higher than the same value for the wing alone. In the theoretical papers [31, 32], the excess value of the lift-curve slope of the wing-body combination above the same value for the isolated wing of the same geometry was also noted. In paper [32] the midwing monoplane scheme has received an increase in the value of the lift-curve slope of the wing-fuselage combination compared with the isolated wing with the same geometry approximately 19%. Theoretical results of the calculation value of the lift-curve slope of the wing-body combination are devoted also in papers [33–35] and book [36]. Woodward in papers [37, 38] investigated the aerodynamic characteristics of wing-fuselage combinations using the panel method. The same panel method was used in the paper [39]. An experimental study of wing-body-tail combinations was performed in the work [40].

This chapter by no means covers all papers on the interference of the wing and fuselage. Author's book [41] and paper [42] contain more detailed bibliography on the problems of the lift of the wing-body combination.

The main purpose of this paper is to give results of solving optimization problems for the values of the lift wing-body configurations and to demonstrate the conformity of computational author's results with the known experimental and theoretical results of other authors.

2. The calculation method of the interference for wing-fuselage combination

The calculation method of the interference for the wing-fuselage combination [42] includes two methods: (1) a discrete vortex method (DVM) for the surface of the wing and (2) 2D potential model of the flow for cross-flow around fuselage [41].

The original three-dimensional problem (**Figure 1**) is divided into two parts. First part is the two-dimensional problem of the flow around the cross section of the fuselage (**Figure 2**), and the second part is a three-dimensional problem for the isolated wing. In the 2D problem, the flow around the cross section of the fuselage adds a pair of discrete point vortices. The added vortices are the consequence of lift on the wing. According to Zhukovsky's theory about the lift of the wing, any lifting surface can be replaced by an equivalent П-shaped vortex; free vortices at low angles of attack lie in the plane of the wing and extend to infinity. In our model, it is

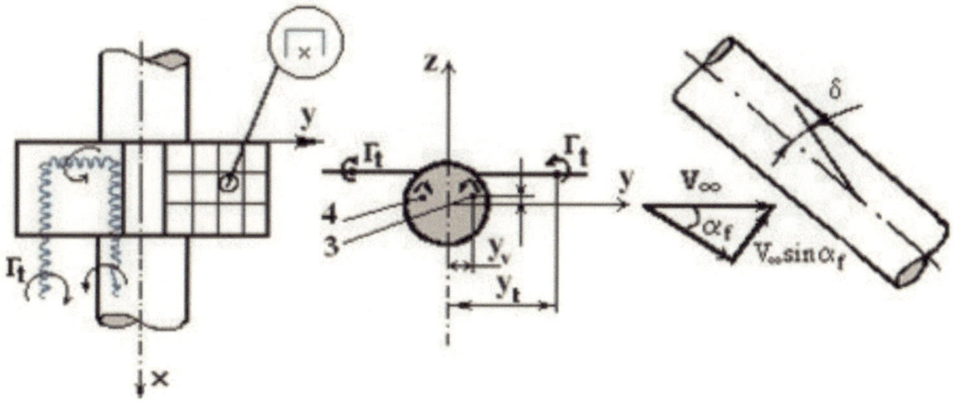

Figure 1.
The mathematical model of wing-body interference.

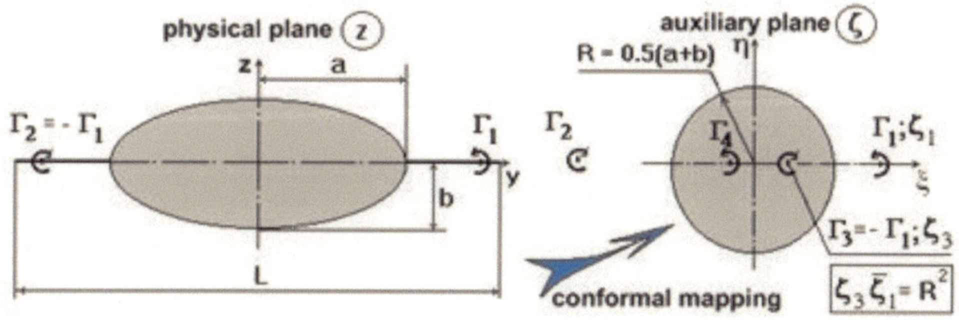

Figure 2.
The mathematical model of the potential flow around the elliptical cross section of the fuselage in the present pair of vortices.

proposed each console part of the wing replaces one Π-shaped vortex lying in the plane of the wing. The Π-shaped vortex in the left-wing console is shown in **Figure 1**. The coordinate of the free vortex and its intensity can be found from the bond equation; after the lift-isolated wing by DVM will be defined. The inversion method (**Figure 2**) can be used to satisfy the boundary conditions of impermeability on the surface of the body cross section for the canonical body, and for the arbitrary two-dimensional cross section can use the panel method. An example solution for the potential flow around the elliptical cross section of the fuselage in the present of the pair vortices is shown in **Figure 3**.

In this formulation, the problem is reduced to solving the following system of algebraic linear equations:

$$\sum_{i=1}^{L} \boldsymbol{\Gamma}_i (\mathrm{A}_{ij} \cdot \mathbf{n}_j) = -(\mathbf{F}_j \cdot \mathbf{n}_j), \quad j = 1, \ldots L, \tag{1}$$

where L is the number of control points (collocation points) equal to the number of attached vortices on the right-wing console, $\mathbf{n}_j$ is the unit normal vector to the jth control point on the surface of the wing, A_{ij} is the matrix of the aerodynamic influence or the matrix of the induced velocities at the control points of the wing surface from all system of horseshoe vortices (left and right consoles for the isolated wing), and $\mathbf{F}_j$ is a column vector of the velocity induced in the jth the control point on the wing surface by incoming flow and the flow from the cross section of the fuselage that includes

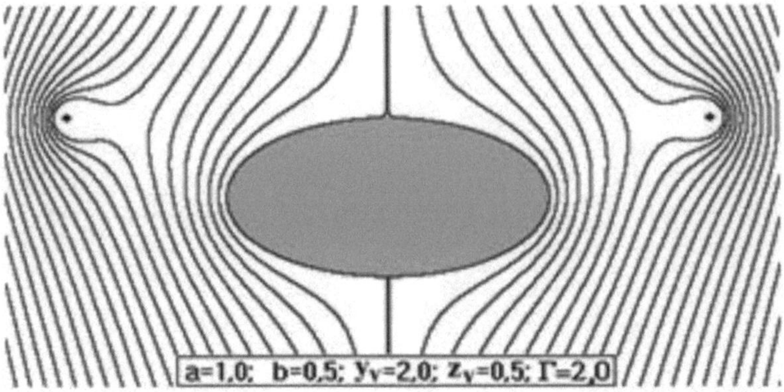

Figure 3.
Streamlines of the potential flow around the elliptical cross section of the fuselage in the present pair of vortices.

either inversion of the vortices or sources and sinks providing satisfying conditions impermeability on the surface body from the free vortices left and right wing.

For small angles of attack of the wing-body combination ($\alpha \ll 1$) and small wing deflection angle (angle of inclination wing), ($\delta \ll 1$) can use the linear formulation, and then a solution can be written as a linear function of the angle of attack and wing deflection angle:

$$\mathbf{\Gamma}_i = \mathbf{\Gamma}_i^{\alpha} \cdot \alpha + \mathbf{\Gamma}_i^{\delta} \cdot \delta. \tag{2}$$

Right parts of the system of algebraic linear equations Eq. (1) can be represented also as

$$\mathbf{F}_j = \mathbf{F}_j^{\alpha} \cdot \alpha + \mathbf{F}_j^{\delta} \cdot \delta. \tag{3}$$

In Eqs. (2) and (3), $\mathbf{\Gamma}_i^{\alpha}$, $\mathbf{\Gamma}_i^{\delta}$, $\mathbf{F}_j^{\alpha}$, $\mathbf{F}_j^{\delta}$ are derivatives of the $\mathbf{\Gamma}_i$, $\mathbf{F}_j$ on the angle of attack α and wing deflection angle δ, respectively.

For calculating the right parts of the system Eq. (1) for the problem with the fuselage of an arbitrary cross section of the body, the panel method that leads to the solution system of algebraic linear equations (4) is proposed:

$$[\mathbf{A}]\ [\boldsymbol{\sigma}] = [\mathbf{R}], \tag{4}$$

where $\boldsymbol{\sigma}$ is a column vector

$$\boldsymbol{\sigma} = \boldsymbol{\sigma}^{\alpha} \cdot \alpha + \boldsymbol{\sigma}^{\delta} \cdot \delta. \tag{5}$$

Let us give the final formula for the components of the induced velocity, for example, for the case of the circular cross section of the fuselage in jth control point of the wing panel:

$$\begin{aligned} V_{nj} &= V_{zj}\cos\delta \\ &= \cos\delta\left\{ V_\infty \sin\alpha\left[1 + \frac{R^2\left(y_j^2 - z_j^2\right)}{\left(y_j^2 + z_j^2\right)^2}\right]\right. \\ &\quad \left. - \frac{\Gamma_t}{2\pi}\left[\frac{y_j - \tilde{y}_v}{\left(y_j - \tilde{y}_v\right)^2 + \left(z_j - \tilde{z}_v\right)^2} - \frac{y_j + \tilde{y}_v}{\left(y_j + \tilde{y}_v\right)^2 + \left(z_j - \tilde{z}_v\right)^2}\right]\right\}, \end{aligned} \tag{6}$$

where $\left(y_j,\ z_j\right)$, $\left(\tilde{y}_v\ \tilde{z}_v\right)$ are coordinates of the control point and the inversion vortex point (see **Figure 1**), respectively, and V_{nj} is a normal component of the velocity to the surface at the jth control point wing panel induced velocity component along the OZ-axis of the cylinder in cross-flow (see **Figure 1**). Coordinate inversion vortices are defined by Milne-Thomson's theorem about the circle [43]. The coordinate y_t and intensity Γ_t of the free vortex can be found on the connection equations [41, 42, 44].

So the task of the wing-body interference is reduced to the solution Eq. (1) with the right part Eq. (6) or right-hand parts, obtained by solving the system (4) that provides the solution of the problem for the potential flow around an arbitrary contour of the panel method. The method of the successive iterations provides an agreement of the velocity field on the surface wing and the surface fuselage. Each iteration is reduced to the solution of systems of linear algebraic equations (1) with corrected right part Eq. (6). The zero iteration can select the solution for the isolated wing. For small angles of attack and wing deflection angle, the proposed model or the linear formulation allows to get the solution of two problems at once, which can be called αα-problem (fuselage and wing have the same angle of attack, angle of the wing deflection angle equal to zero) and δ0-problem (the fuselage has a zero angle of attack, and the wing has deflection angle δ not equal to zero). For this linear case, formulas for the coefficients of the normal forces of the wing and the body are of the form Eq. (7)

$$\begin{aligned}
&C_{L\ \mathrm{W(B)}} = C_{L_\alpha\ \mathrm{W(B)}}\alpha + C_{L_\delta\ \mathrm{W(B)}}\delta,\\
&C_{L\ \mathrm{B(W)}} = C_{L_\alpha\ \mathrm{B(W)}}\alpha + C_{L_\delta\ \mathrm{B(W)}}\delta,\\
&C_{L_\alpha\ \mathrm{W,\,B}} = C_{L_\alpha\ \mathrm{W(B)}} + C_{L_\alpha\ \mathrm{B(W)}},\quad C_{L_\delta\ \mathrm{W,\,B}} = C_{L_\delta\ \mathrm{W(B)}} + C_{L_\delta\ \mathrm{B(W)}},
\end{aligned} \tag{7}$$

where values $C_{L_\delta\ W(B)}$, $C_{L_\delta\ B(W)}$, $C_{L_\alpha\ W(B)}$, $C_{L_\alpha\ B(W)}$ are obtained from the solution δ0- and αα-problem, respectively.

3. Calculation results

The comparison of the calculation results obtained from the above theoretical model with calculations by the DVM for case αα-problem [34–36] is shown in **Figures 4–9**. The rectangular, triangular, and swept wings were considered. It may be noted is enough good agreement of calculated data.

The changing of coordinates of the aerodynamic center x_{AC}/c measured from the beginning of the mean aerodynamic chord is also shown in **Figures 4**, **6**, and **8**. The coefficient of the interference K_Σ in these figures is defined by the formula

$$K_\Sigma = \frac{C_{L_\alpha \mathrm{W,\,B}}}{C_{L_\alpha \mathrm{W}}}, \tag{8}$$

where value $C_{L_\alpha \mathrm{W}}$ is a lift-curve slope for an isolated wing composed of two consoles of this wing.

The comparison of the calculation results obtained from the above theoretical model with calculations by the numerical method of singularities for case δ0-problem [31] is presented in **Figure 10**.

The comparison of calculated data for the mathematical model described above and the calculated and experimental data of other researchers [45–49] is shown in **Figures 11–13**.

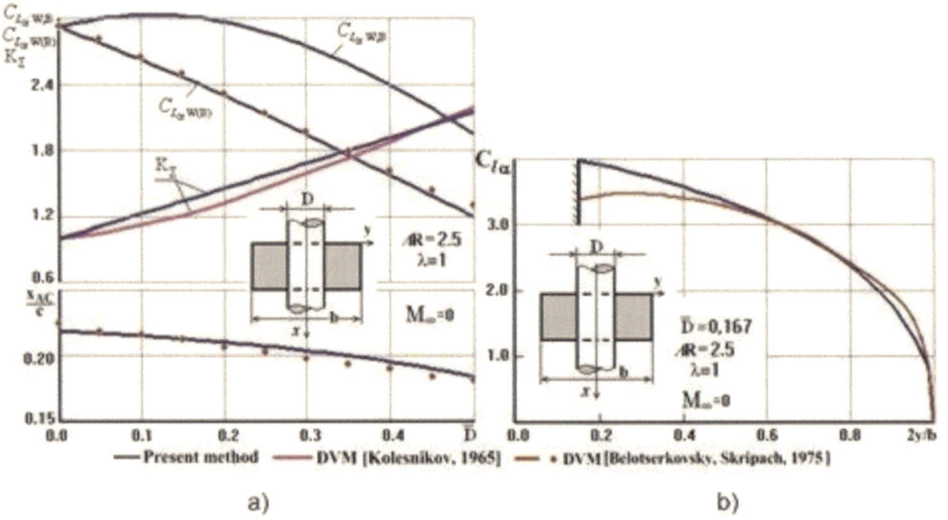

Figure 4.
The lift-curve slopes vs. relative diameter of the fuselage (a); the lift-curve slopes vs. relative span for the rectangular wing in the midwing-body combination (b).

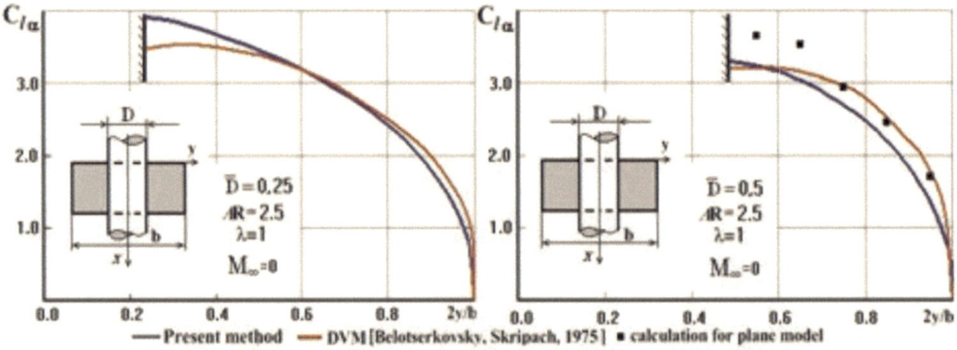

Figure 5.
The lift-curve slopes vs. relative span for the rectangular wing in the midwing-body combination.

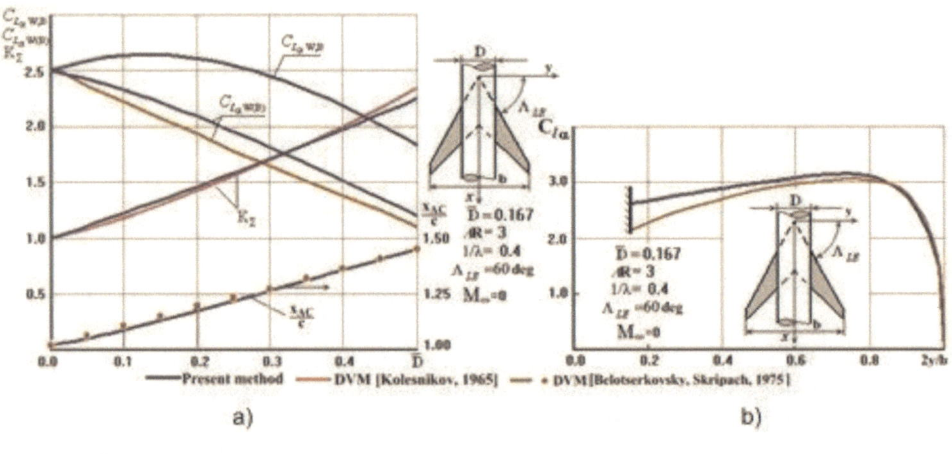

Figure 6.
The lift-curve slopes vs. relative diameter of the fuselage (a); the lift-curve slopes vs. relative span for the swept wing in the midwing-body combination (b).

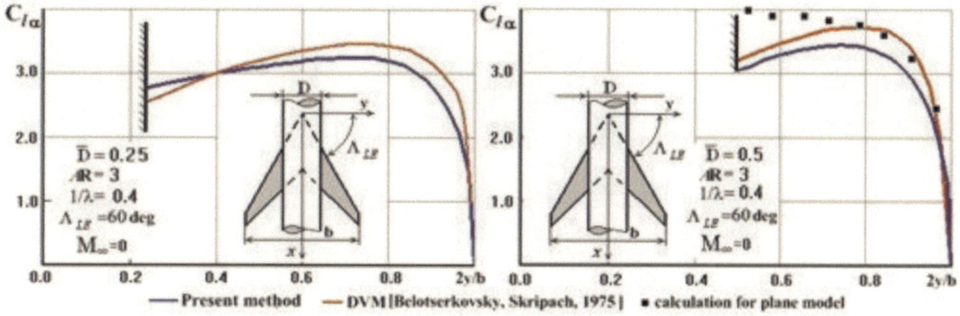

Figure 7.
The lift-curve slopes vs. relative span for the swept wing in the midwing-body combination.

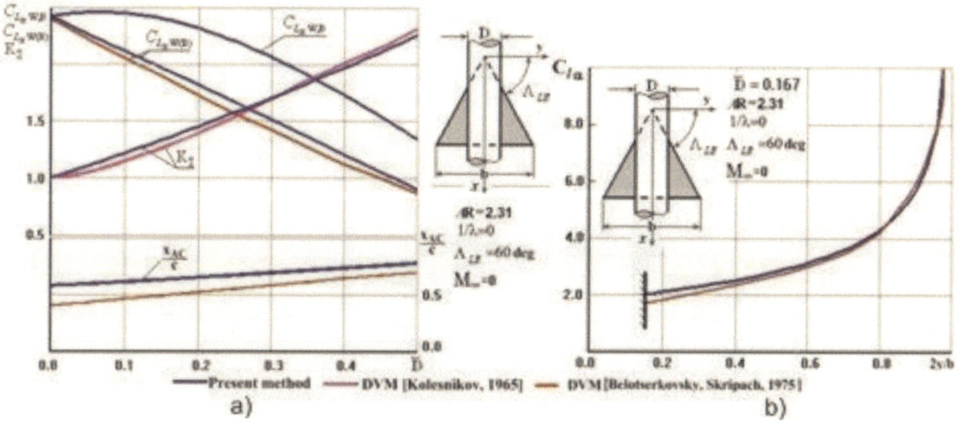

Figure 8.
The lift-curve slopes vs. relative diameter of the fuselage (a); the lift-curve slopes vs. relative span for the delta-shaped wing in the midwing-body combination (b).

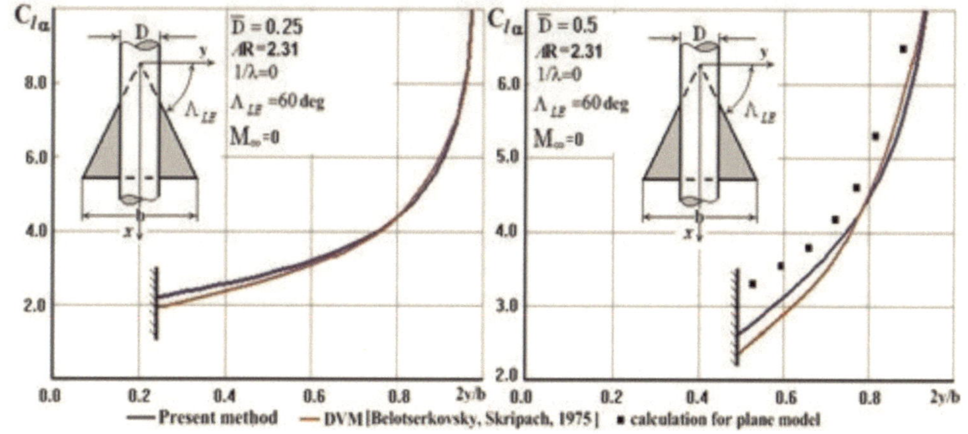

Figure 9.
The lift-curve slopes vs. relative span for delta-shaped in the midwing-body combination.

Figure 14 shows an influence of compressibility on the values of theoretical lift-curve slopes for case midwing monoplane combination with the rectangular and delta-shaped wing. **Figure 15** also shows an influence of compressibility on values of theoretical lift-curve slopes for case high-wing monoplane combination with the rectangular and delta-shaped wing [33].

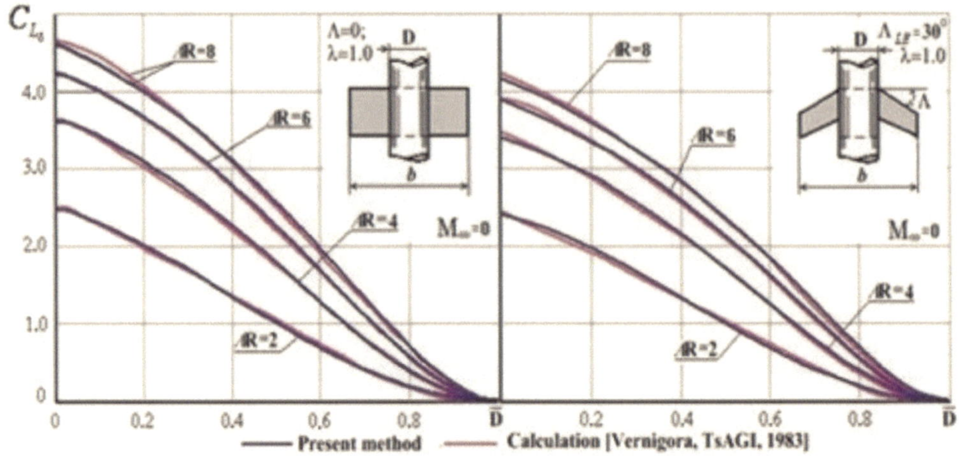

Figure 10.
The lift-curve slopes vs. relative diameter of the fuselage for case δo-problem for midwing-body combination.

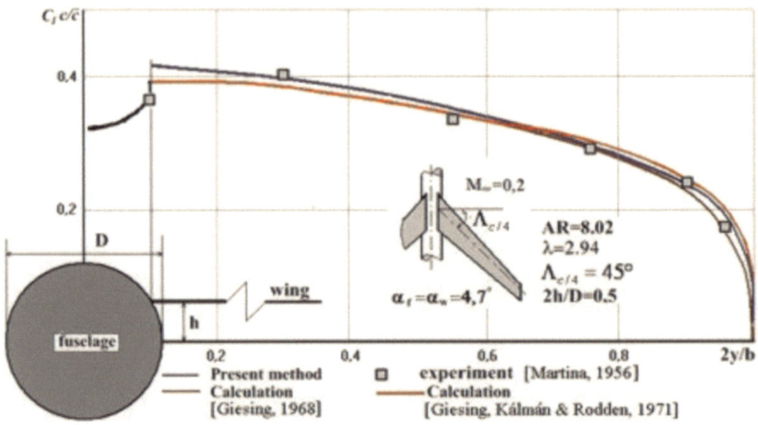

Figure 11.
The distribution lift coefficient along the relative span of the swept wing for case high-wing monoplane combination.

Of particular interest is the comparison of calculated and experimental data to prove that the lift-curve slope for the wing-body combination exceeds this value for an isolated wing. **Figure 16** shows this comparison.

The area shown in color in **Figure 16** indicates the advantage of the lift-curve slopes of the wing-body combinations over an isolated wing. Calculations and experiments show that with an increasing aspect ratio of the wing, this advantage will increase. This circumstance is important since the modern development of the aircraft industry tends to increase the aspect ratio of the wing. Another conclusion is that the maximum of the lift-curve slopes with a wing aspect ratio of 6 is achieved at relative fuselage diameter of approximately 0.2. Such a relative diameter of the fuselage allows the design of modern aircraft with a wide fuselage. Numerical studies have shown that with increasing aspect ratio of the wing and the ratio of the width to the height of the fuselage elliptical cross sections, the advantage of lift-curve slopes of the wing-body combinations over isolated wings becomes larger. The noted facts allow us to formulate and solve an optimization problem.

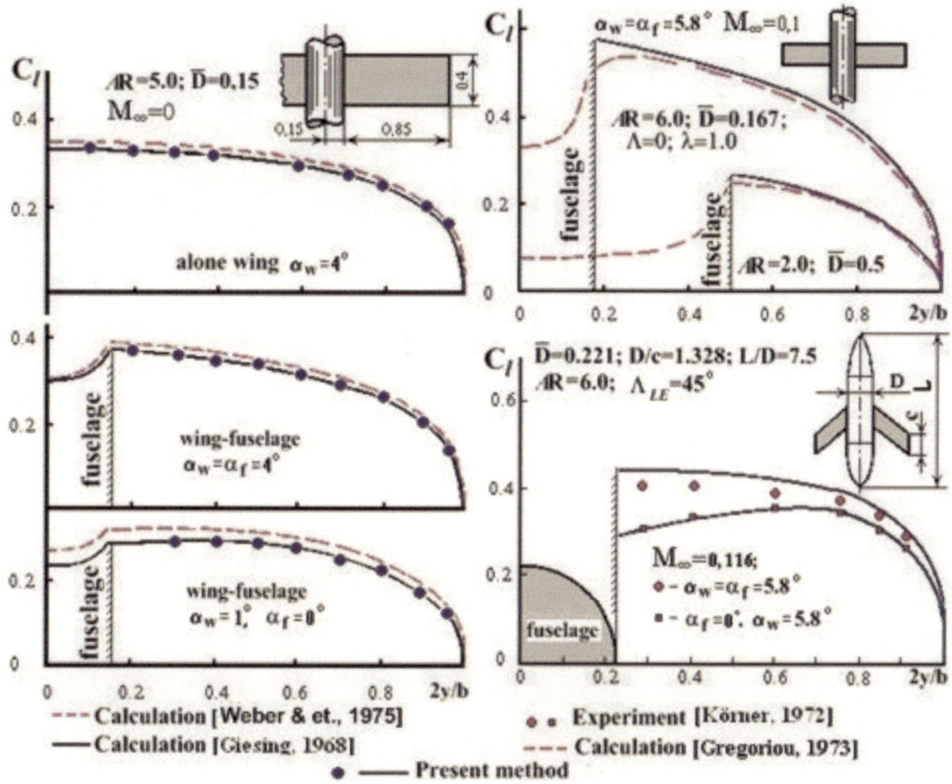

Figure 12.
The distribution lift coefficient along the relative span of the rectangular wing for case high-wing monoplane combination.

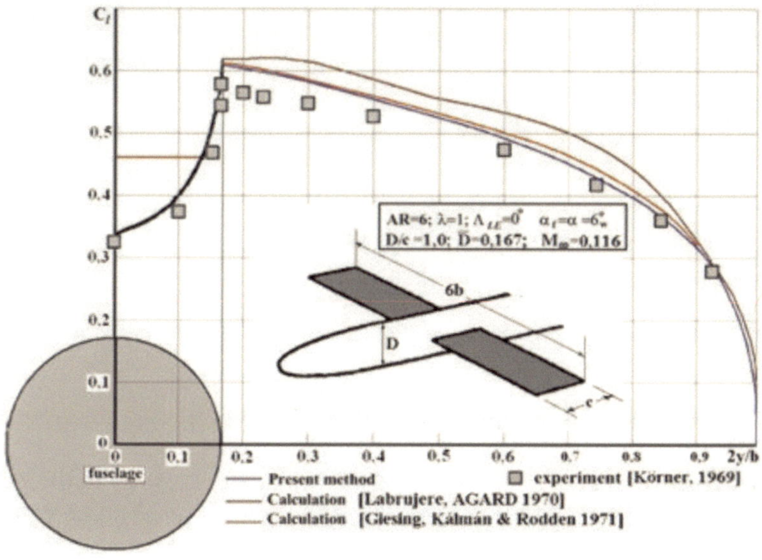

Figure 13.
The distribution lift coefficient along the relative span of the rectangular wing for case midwing monoplane combination.

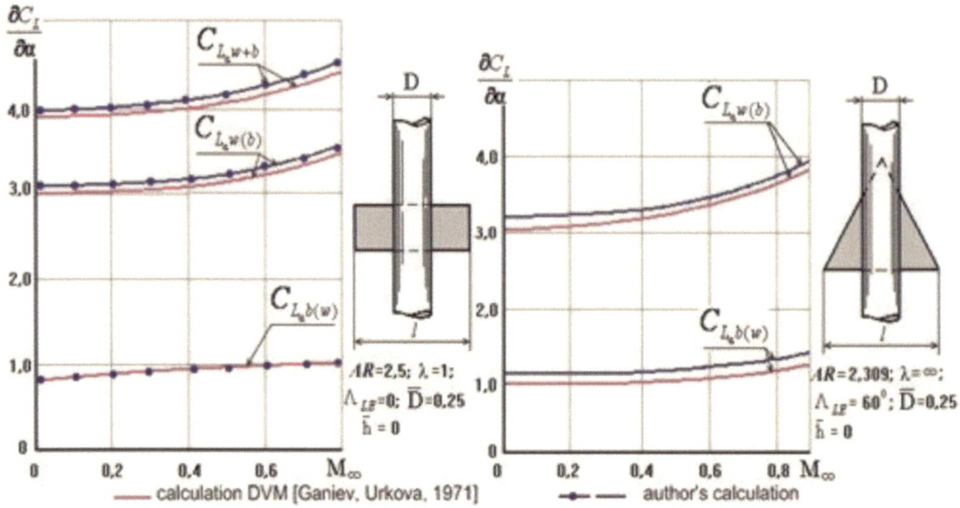

Figure 14.
Calculation results of lift-curve slopes vs. Mach number for case midwing monoplane combination.

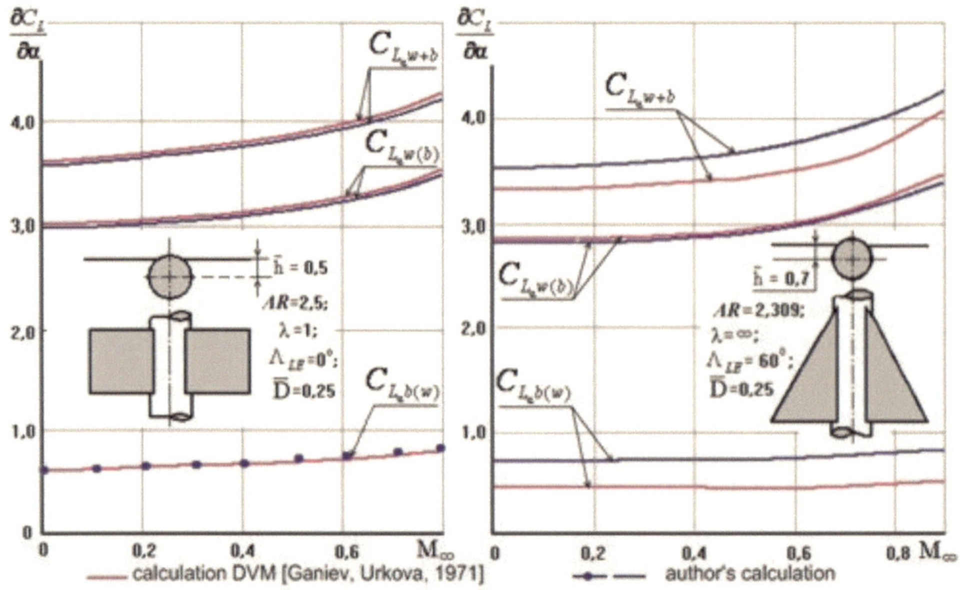

Figure 15.
Calculation results of lift-curve slopes vs. Mach number for case high-wing monoplane combination.

4. The formulation of the optimization problem

Note that in some theoretical and experimental papers devoted to the wing-body interference revealed a maximum dependence $\partial C_L/\partial\alpha = f(d_f/b)$. Our calculations on the above mathematical model also confirm this fact. It was found that the maximums of lift-curve slopes for a wing-body combination depends on the shape of the wing and the cross-section shape of the fuselage. The paper presents solutions to the optimization problem for the wing-body combinations with unswept trapezoidal wings and circular or elliptical cross sections.

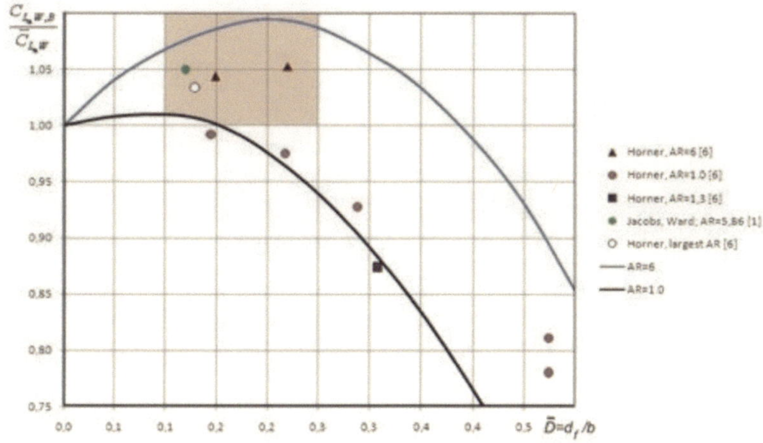

Figure 16.
Theoretical and experimental results for lift-curve slopes vs. relative diameter of the fuselage.

We will use the formulation of the optimization problem as a nonlinear programming problem as follows:

$$max \quad C_{L_\alpha W,\,B}(\mathbf{X}), \quad \mathbf{X} \in E^n, \tag{9}$$

where $\mathbf{X} = [x_1, x_2, 0]^{\mathrm{T}}$ is a vector of the project parameters connected with geometrical characteristics of the wing-body configuration by formulates

$$\overline{D} = \frac{d_f}{b} = \frac{1}{x_1^2 + 1}, \quad \frac{1}{\lambda} = \frac{1}{x_2^2 + 1}, \tag{10}$$

where $x_1 x_2$ are auxiliary variables. The problems Eq. (9) and (Eq. (10) are a problem of unconditional optimization, for which there are $\overline{D} \in [0;1]$, $(1/\lambda) \in [0;1]$ and $x_1 \in [-\infty; +\infty]$, $x_2 \in [-\infty; +\infty]$.

5. Results of the optimization problem for lift-curve slope for midwing-body monoplane configuration

Figures 17 and **18** show results of the optimization problem for lift-curve slope for midwing-body monoplane configuration with circular cross-section fuselage vs. the aspect ratio of the rectangle wing. In **Figure 16**, the notation is used:

$$\overline{K}_\Sigma = \frac{C_{L_\alpha} W,\,B}{\overline{C}_{L_\alpha W}},$$

where $C_{L_\alpha} W,\,B$ is the lift-curve slope of the wing-body combination the same as in Eq. (9) and $\overline{C}_{L_\alpha W}$ is a lift-curve slope of the isolated wing in which it is included part of the occupied fuselage.

Figure 18 shows the results of the solution of the optimization problem for lift-curve slopes for midwing-body monoplane configuration with elliptical cross-section fuselage. Maximum values of the lift-curve slopes depend on the aspect ratio of the rectangular wing and the ratio of the axes of the ellipse. **Figure 18** shows that the advantage of the wing-fuselage combination over an isolated wing is enhanced

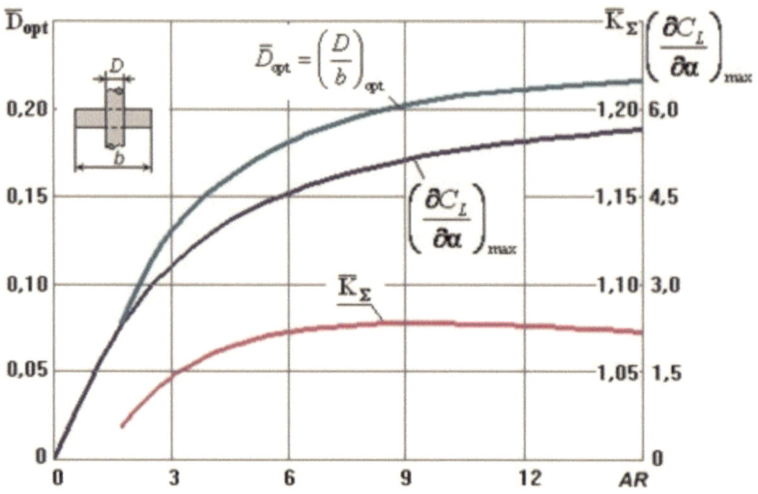

Figure 17.
The optimal relative diameter of the fuselage with circular cross-section body for the midwing configuration vs. the aspect ratio of the rectangular wing.

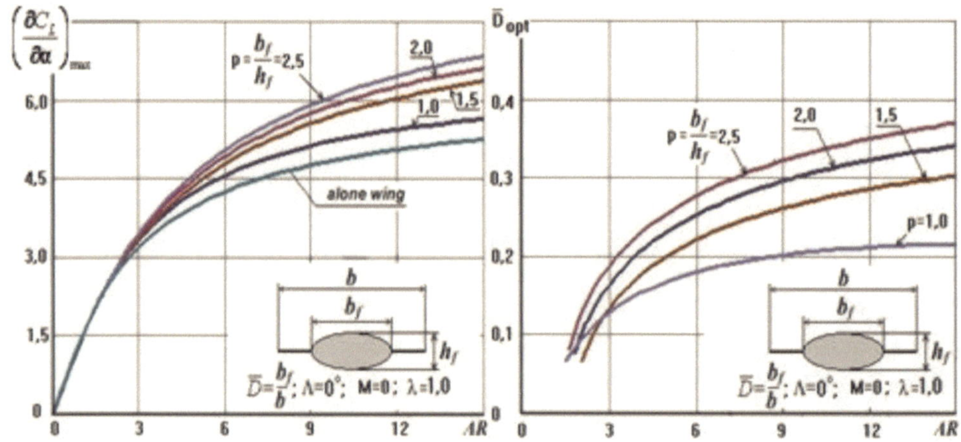

Figure 18.
Maximums of the lift-curve slopes for the optimal midwing configuration with elliptical cross-section body vs. the aspect ratio of the rectangular wing.

with increasing the aspect ratio of the rectangular wing and with increasing the ratio of the axes of the cross-section fuselage. The optimal ratio of the width of the body to the span of the wing can reach 30% and more!

Figure 19a shows the effect of the compressibility and the statistics for modern aircraft also (**Figure 19b**). Red color point shows the project of fifth-generation aircraft (project M-60, Russia). The feature of the project M-60 is a wide fuselage. As can be seen from **Figure 19b**, with the aspect ratio wing equal to 15, the optimal ratio of the width of the circular cross section to the wingspan can reach 20%!

6. Conclusions

The paper presents results obtained by the author for wing-body interference. The lift-curve slopes of the wing-body combinations are considered. A 2D potential

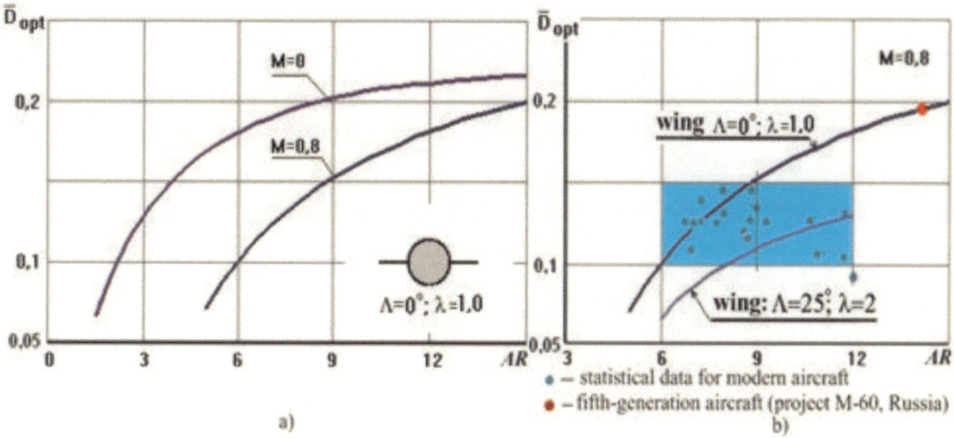

Figure 19.
Effect compressibility on the solution of the optimization problem.

model for cross-flow around the fuselage and the discrete vortex method for the wing were used. Flat wings of various forms and the circular and elliptical cross sections of the fuselage are considered. It was found that the value of the lift-curve slopes of the wing-body combinations may exceed the same value for an isolated wing. An experimental and theoretical data obtained by other authors earlier also confirms this result. Investigations to optimize the wing-body combination were carried within the framework of the proposed model. The proposed mathematical model for the solution optimization problem for the wing-body combination allows selecting the optimal geometric parameters for configuration to maximize the values of the lift-curve slopes of the wing-body combination.

It was revealed that the maximums of the lift-curve slopes for the optimal midwing configuration with elliptical cross-section body reach their values at sufficiently large relative width of the body (more than 30% of the span wing!). The advantage of the wing-fuselage combination with a circular cross section over an isolated wing at the wing aspect ratio greater than 6 can reach 7.5% at the relative diameter of fuselage equal to approximately 0.2. The advantage of the wing-fuselage combination with the elliptical cross section with the ratio of axes of the body equal to 2.5 over an isolated wing with aspect ratio equal to 12 is that it can reach 29% at relative width of fuselage equal approximately to 0.35!

Conflict of interest

The author declares no conflict of interest.

Acronyms and abbreviations

NACA	National Advisory Committee for Aeronautics
AGARD	Advisory Group for Aerospace Research and Development
ZFF	Zeitschrift für Flugwissenschaften
TsAGI	Central Aerohydrodynamic Institute
J. Aircraft	Journal of Aircraft
R&M	Reports and Memoranda
J. Aeron. Sc.	Journal of the Aeronautical Sciences
ARC CP	Aeronautical Research Council Current Papers

Author details

Vladimir Frolov
Samara National Research University, Samara, Russia

*Address all correspondence to: frolov_va_ssau@mail.ru

References

[1] Jacobs EN, Ward KE. Interference of wing and fuselage from tests of 209 combinations in the N.A.C.A. Variable Density Tunnel. NACA Report; 1935. 540, p. 37

[2] Pitts WC, Nielsen JN, Kaattari GE. Lift and center pressure of wing-body-tail combinations at subsonic, transonic and supersonic speeds. NACA Report; 1957. 1307, p. 70

[3] Schneider W. Experimental investigation of wing-body interferences in the mach number range from 0.5 to 2.0. Transonic Aerodynamics. AGARD CP No. 35; Göttingen, Germany: Aerodynamische Versuchsanstalt Göttingen; 1968. pp. 20-1-20-23

[4] Kirby DA, Hepworth AG. Low-speed wind-tunnel tests on some slender airbus configurations. R&M; 1971. No. 3747, p. 55

[5] Körner H. Berechnung der potentialtheoretischen Strömung um Flügel-Rumpf-Kombinationen und Vergleich mit Messungen. ZFF; Vol. 20. 1972. pp. 351-368

[6] Korner SF, Borst HV. Fluid-Dynamic Lift. 2d ed. Published by Mrs. L.A; 1985. Hoerner, 482 p

[7] Golubinsky AI. Exact solution of interference problem of wing with fuselage at subsonic flow. In: Proceedings TsAGI; 1961. 810, pp. 23-36

[8] Lebedev BF. Approximate method of calculating load distribution on wing and fuselage at subsonic speeds. In: Proceedings TsAGI; 1958. p. 719

[9] Dorodnitsyn AA. Influence of fuselage on distribution of loads on wing span. In: Proceedings TsAGI; 1944. p. 546

[10] Keldysh VV. Interference of flat swept low-aspect ratio wing and body. In: Proceedings TsAGI; 1959. 759, pp. 1-23

[11] Keldysh VV. Lift small aspect ratio wing with body. In: Scientists Note TsAGI; VI, 5; 1975. pp. 15-28

[12] Keldysh VV. Lift and longitudinal moment of low-aspect ratio wing with body of rotation located near it. In: Scientists Note TsAGI; VIII, 3; 1977. pp. 19-31

[13] Flax AH. Comment on "correlation of wing-body combination lift data". Journal of Aircraft. 1974;**11**(5):393-394

[14] Flax AH. Integral relations in linearized theory of wing-body interference. Journal of the Aeronautical Sciences. 1953;**20**(7):483-490

[15] Flax AH, Lawrence HR. The aerodynamics of low-aspect-ratio wings and wing-body combinations. In: Third Anglo-American Aeron. Conf. Brighton; 1951; pp. 363-398, 398A-398J

[16] Lawrence HR, Flax AH. Wing-body interference at subsonic and supersonic speeds—Survey and new development. Journal of the Aeronautical Sciences. 1954;**21**(5):289-324

[17] Nielsen JN. Missile Aerodynamics. New York-Toronto-London: McGraw-Hill Book Company, Inc.; 1960. 450 p

[18] Spreiter JR. Aerodynamic properties of slender wing-body combination at subsonic, transonic and supersonic speeds. NACA TN; 1948. 1662

[19] Spreiter JR. Aerodynamic properties of the cruciform wing and body combinations at subsonic, transonic, and supersonic speeds. NACA TN; 1949. 1897

[20] Spreiter JR. Aerodynamic forces on slender plane and cruciform wing and body combinations, NACA Report. 1950. 962, pp. 271-287

[21] Spreiter JR, Sacks AH. A theoretical study of the aerodynamics of slender cruciform-wing arrangements and their wakes. NACA Report; 1956. 1296, pp. 81-106

[22] Ferrari C. Aerodynamic components of aircraft at high speeds vol. VII. In: Donovan AF, Lawrence HR, editors. High-Speed Aerodynamics and Jet Propulsion. Princeton, New Jersey: Princeton University Press; 1957. pp. 228-435

[23] Karafoli E. Aerodynamics of an Aircraft Wing. Moscow: AN USSR Publishing; 1956. 479 p. (In Russian)

[24] Ryzhkov UA. Approximate method of calculating forces acting on body in supersonic gas flow and problem of aerodynamic interference of wing-fuselage. In: Proceedings TsAGI; 1959. 759, pp. 24-37

[25] Lennertz J. On the mutual reaction of wings and body. NACA TM, 400; 1927

[26] Lennertz J. Beitrag zur theoretischen Behandlung des gegenseitigen Einflusses von Tragfläche und Rumpf. Abhandlungen aus dem Aerodyn. In: Durand WF, editor. Aerodynamics Vol. IV. Berlin: Springer; 1934

[27] Nikolsky AA. About lifting properties and induced drag of wing-fuselage system. In: Proceedings TsAGI; 1981. 2122, pp. 94-95

[28] Weber J. Interference problems on wing-fuselage combinations. Part I: Lifting unswept wing attached to a cylindrical fuselage at zero incidences in midwing position. RAE Technical Report 69130 (ARC 31532); 1969

[29] Weber J. Second-order small-perturbation theory for finite wings in incompressible flow. ARC R&M 3759; 1972

[30] Weber J, Joyce MG. Interference problems on wing-fuselage combinations Part III symmetrical swept wing at zero incidence attached to a cylindrical fuselage. ARC CP; 1975. 1333, p. 84

[31] Vernigora VN. Numerical study of interference of thin trapezoidal wing and cylindrical fuselage at subsonic speeds. In: Proceedings TsAGI; 1983. 2176, pp. 26-43

[32] Kholyavka VI. Interference of aircraft parts. Kharkiv: Aviation. Inst.; 1967. p. 107

[33] Ganiev FI, Urkova ME. Calculation results of aerodynamic characteristics of wing-body combination moving with high subsonic speed. In: Proceedings Air Force Academy named after Zhukovsky; 1971. 1302, pp. 220-241

[34] Kolesnikov GA. Calculation of aerodynamic characteristics of wing-body combination at unsteady motion. In: Proceedings TsAGI; 1961. 810, pp. 78-96

[35] Kolesnikov GA. Order and results of calculation of aerodynamic characteristics of wing-body combinations at unsteady motion. In: Proceedings TsAGI; 1965. 954, pp. 3-28

[36] Belotserkovsky SM, Skripach BK. Aerodynamic Derivatives Wing and the Aircraft at Subsonic Speeds. Moscow: Nauka; 1975. 424 p. (In Russian)

[37] Woodward FA. Analysis and design of wing-body combinations at subsonic and supersonic speeds. Journal of Aircraft. 1968;5(5):528-534

[38] Woodward FA. An improved method for the aerodynamic analysis of

wing-body-tail configuration in subsonic and supersonic flow. NASA CR-2228, Part I and Part II; 1973. pp. 126 and 315

[39] Cvitanović I, Virag Z, Krizmanić S. Analysis of potential flow around wing-body configurations. In: 4th International Congress of Croatian Society of Mechanics (4th ICCSM); 18-20 September 2003; Bizovac, Croatia; 2003. pp. 1-10

[40] Davari AR, Soltani MR, Askari F, Pajuhande HR. Effects of wing geometry on wing-body-tail interference in subsonic flow. Scientia Iranica. 2011;**18**(3):407-415

[41] Frolov VA. Methods for Calculating Lift for Wing-Body Combination. Saarbrucken: LAP Lambert Academic Publishing; 2011. 141 p. (in Russian)

[42] Frolov VA. Review results on wing-body. In: Proceedings of the International Conference on Mechanical, System and Control Engineering (ICMSE 2016); 18-20 May 2016; Moscow, Russia: MATEC Web of Conferences ICMIE; 2016. 75, 09006. DOI: 10.1051/matecconf/2016750900

[43] Milne-Thomson LM. Theoretical Aerodynamics. London: Constable and Company; 1958

[44] Mendenhall MR, Nielsen JN. Effect of symmetrical vortex shedding on the longitudinal aerodynamic characteristics of wing-body-tail combinations. NASA CR, 2473; 1975. p. 115

[45] Giesing JP, McDonnell-Douglas Rep. Douglas Aircraft Co., Report DAC, Vol. I, 67212. 1968

[46] Giesing JP, Kalman WP, Rodden WP. Subsonic unsteady aerodynamics for general configurations; Part I—Direct application of the nonplanar doublet-lattice method; Part II—Application of the doublet-lattice method and the method of images to lifting surface/body interference. USAF FDL-TR-71-5; 1971

[47] Martina AP. The interference effect of a body on the spanwise load distribution of two 45° sweptback wings of aspect ratio 8.02 from low-speed test NACA TN. 1956; p. 3730

[48] Gregoriou G. Zur gegenseitigen Beeinflussung eines endlich Langen Rumpfes und eines Flügels in Mitteldeckeranordnung bei Unterschallströmung. BMVg-FBWT-73-33; 1973

[49] Labrujere TE, Loeve W, Slooff JW. An approximate method for the calculation of pressure distribution on wing-body combinations at subcritical speeds. AGARD-CP, No. 071; 1970

Section 3

Boundary-Layer Theory

Chapter 5

Wave Drag Modification in the Presence of Discharges

Anna Markhotok

Abstract

A phenomenological model of wave drag modification following the thermal energy deposition in a hypersonic flow is presented. While most of the previous research was concentrated on finding optimal gas parameter values and the amount of energy, this work points at closer attention to the effect of the parameter distribution and the geometry of experimental arrangements. The approach discussed here is to fill the gap in the understanding of the complex mechanism of the flow transformation leading to the wave drag reduction. Analytical expressions used in the model identify a number of adjustment parameters that can be used to optimize thermal energy input and thus achieve fundamentally lower drag values than that of conventional approaches.

Keywords: shock waves, hypersonic plasma dynamics, shock wave instability, vorticity, aerodynamics drag

1. Introduction

The drag is one of the four major forces acting on an airplane in flight. Depending on the nature of the body-flow interaction and the phase of flight, it can be divided into three basic components. The pressure drag is due to pressure distribution over the vehicle surface and skin friction. The induced drag results from the pressure redistribution following the formation of the trailing vortex system during the lift production and thus becomes significant only during takeoff and landing times. When the free stream Mach number exceeds 0.8, the local flow velocities at some points on the airframe may become supersonic, and shock waves set in at the corresponding locations. The gas compression in a shock wave results in the pressure increase in the flow in front of the moving body. This is usually felt as an abrupt increase in local drag and reduction in local lift that may be accompanied by associated changes in trim, pitching moment, and the stability and control characteristics of the airplane. This portion of the drag, the so-called wave drag, is up to four times larger than its subsonic component and can result in the aerodynamic efficiency reduced by more than 50%. The wave drag is the reason for the strong pushback (shock stall) an aircraft experiences when, during acceleration through the transonic flight regime, the Mach number first reaches its critical value and the local speed passes through the sonic point ("sonic barrier"). The increase in the drag by 1% is equivalent to 5–10% of payload, and thus it imposes significant limitations on the speed, range, and payload carried by a vehicle [1].

The shock wave weakening achieved by altering the shock formation processes is one of the efficient methods developed to reduce the wave drag. Swept wing surface minimizes the shock strength, thereby reducing wave drag. Sweeping beyond the Mach line converts the wing design problem into a transonic issue and also enables extension into supersonic speeds of "leading edge thrust" [1, 2]. Moving lift forward onto the fuselage allows extending the lifting line and reducing the drag due to lift [3]. This approach finally led to the so-called Coke bottle of the transonic fighter regime [4]. A method using an active or passive flow separation control at cruise has been suggested in [5].

The shock wave interaction with other components of the aircraft (centerbody-nacelle, nacelle-wing, body-wing, body-body, and body/wing-propulsion exhaust efflux) [1, 4] can be used to obtain favorable wave interference leading to a weakened shock system. Up to 25% reduction of the wave drag and increase in the lift-to-drag ratio were reported [6].

One of the approaches in which the shock energy is effectively redirected into the thrust production uses the interaction between the forebody shock with a wing raised above the body. The shock reflection onto the afterbody results in pressure increase and an increment of thrust. Though these methods were proven to be effective in a number of localized problems such as centerbody inlets and nacelle-wing interactions, significant difficulties were encountered in other applications [1].

The local passive porosity method [7] utilizes high transonic flow sensitivity to local geometry changes resulting in the replacement of strong shocks with a system of weaker ones (λ-shock system). Localized wall deformations/bumps were shown to weaken the shock wave through the generation of nearly isentropic waves upstream [8]. The trailing edge methods use truncation of the region or morphing trailing edge to alter the airfoil shape [9].

The "physical spikes" representing an artificial sharpening of the front region of a blunt body can reduce the drag up to 30–45% [10]. It works through the replacement of nearly normal shock wave with a weaker, oblique shock structure. The upstream inert gas jets or liquid/solid particles injected in front of the body [11] is another method of the same type; however, its effect is weakened by injection thrust in the drag direction.

Thermal energy deposited in the flow upstream of the body is another spike-like working approach reporting 20–30% drag reduction rates in experiments [5, 12, 13] and 40–96% (with 65-fold "return" of the invested energy) predicted in numerical studies [14]. The energy is delivered into the flow via RF, microwave, or optical discharge, with focused electron beam or localized combustion combined with the spike. Pulsed energy deposition methods have been shown to be more efficient than continuous if an appropriate repetition rate is chosen [15]. A combination of methods, such as simultaneous addition of the heat at the tip of the physical spike, leads to flow reattachment on the nose and can further reduce the drag up to 75% [1, 5]. The study showed that, under a number of simplifying assumptions, a thermally created phantom body enveloping the airplane results in finite rise-time signatures that can theoretically eliminate the shock wave, but practically this would require a power input approximately equivalent to twice that necessary to sustain the airplane flight. Thus, the idea of the thermal energy deposition in the flow can be practically realized if highly effective ways of the energy deposition and the shock-thermal area interaction are found.

Using different approaches in the heat design, some authors found little or no effect in the drag reduction. It is the main conclusion of many researchers that, for further progress, a careful design of the energy deposition must be applied. While most of the previous research was concentrated on finding optimal gas parameter values and the amount of energy, this work will point at the exact mechanism of the shock-plasma interaction. Recommendations in the form of adjustment parameters will be done based on closer attention to the effect of the gas parameter distribution and the geometry of the interaction zone. A model providing an insight on the origin of the chain of consecutive transformations in the flow leading to the wave drag reduction can be used to optimize thermal energy input and thus achieve fundamentally lower drag values than that of conventional approaches.

In the interaction between a shock wave and a plasma region produced in a discharge, the initially simply structured planar, bow, or oblique shock wave loses its stability and eventually evolves into a complicated system of distorted and secondary shocks with flow separation regions and formation of vortices in the flow behind the shock [16, 17]. The main changes in the "shock-plasma" system include the shock's acceleration and its front distortion gradually increasing with time, followed with its weakening until it becomes unidentifiable [18, 19]; motion of the shock away from the body in the presence of heating [20]; substantial gas/plasma parameter redistribution in the flow behind the shock, particularly considerable reduction of gas pressure; a vortex system forming in the aftershock flow [12]; remarkable positive dynamics in the vortex strength as the shock propagates well beyond the interface suggesting not only boundary but volume effects as well; the time delays in the effects on the flow relative to the discharge on-off times and a finite pressure rise time [21]; and finally a strong distortion or collapse of the plasma region [22]. Similar phenomena are commonly observed in other settings involving the shock interacting with an interface: in the front separation region control experiments [23], in combustion [22], and in impulsively loaded flows having place in astrophysics plasmas [24] and fusion research [25]. The shock-flame interaction, where the instability induced by the shock wave passage through a flame results in a sharp increase of the burning velocity [26, 27], is used in the supersonic combustion for increased chemical reaction rates.

The existing models used to describe the observed phenomena can be split into two major approaches. The first one considering thermal heating as a possible reason includes Mach number decrease due to gas heating and the possibility of mean molecular weight and number density change caused by molecule dissociation and ionization [14]. The contribution of possible nonthermal mechanisms of the interaction is considered for flows involving atomic and molecular transitions, gas kinetics, electrical properties of plasmas, and non-equilibrium states as a result of fast evolving processes such as radiation or fast expansion. Among them are appearance of charged particles leading to upstream momentum transfer in the hypersonic flow [12, 28, 29]; the possibility of deflection of the incoming flow by plasma in front of the shock via electronic momentum transfer collisions [30]; and the release of heat into the shock layer by the exothermic reactions enhancing the shock layer temperature and thus reducing the pressure and the density behind the shock wave [31].

The baroclinic effect is one of the mechanisms that can be responsible for vorticity generation on an interface when there is nonalignment of pressure and density gradients in the hot gas region [32]. Richtmyer-Meshkov instability (RMI) takes place when two gases of different densities are accelerated by a passage of a shock wave. The acceleration causes growth of small perturbations of the interface, followed by a nonlinear regime with bubbles appearing in the case of lighter gas penetrating a heavier gas and with spikes appearing in the case of a heavier gas moving into the lighter one. The vortex sheet rolls up and accumulates into periodic vortex cores in the post-shock flow [21].

The previous research on this topic was quite successful in explaining some separate features of the interaction. However, the experiments show that each of the processes is not an isolated phenomenon and their coexistence and specific time sequence in their development cannot be covered by the existing models. The model described below is capable of explaining the full set of the features observed in experiments and thus fills the gap in understanding of this phenomenon. The shock refraction on an interface will be considered there as a mechanism [33] that triggers the chain of subsequent flow transformations leading to the wave drag reduction. The model has an advantage of pointing at the origin of the complex phenomena and describing each of the consecutive stages of its development in adequate timing order.

2. Model of the interaction

It will be assumed here that a thermal spot (plasma region) has been created via a discharge at a distance in front of a stationary solid body and the spot's continuous expanding ceases by the time of interaction. This could be a case of a relatively slow evolving plasma bubble or when thermal energy is deposited far enough from the body, so the system had enough time to achieve a thermal equilibrium state. The plasma region is allowed to move with the hypersonic flow toward the bow shock formed in front of the body. The interaction starts when the plasma region arrives at the shock location, and that is when the time t in the model relations starts to be counted. For blunt bodies that are common in the experiments, a spherical shock wave and interface geometry can be an adequate approximation.

In **Figure 1**, the spherical interface (dashed curve) separates the hot plasma (medium 2) and the surrounding cold gas (medium 1). Both media will be treated as ideal gases with initially equal pressures on both sides of the interface.

The cold gas temperature T_1 will be always distributed homogeneously, and T_2 is the plasma temperature right behind the interface with varied parameter distribution along the shock motion direction. The gas temperature can change across the interface abruptly (stepwise) or smoothly (distributed over a distance) [34], and $T_2 > T_1$. The radii R_b and R_s are for the plasma boundary and the shock front correspondingly, and other parameters are shown in the figure. In the reference frame stationary for the plasma region, the shock wave is incident on the interface from left to right, center to center, with constant horizontal velocity V_1. If the hotter medium (plasma) is uniform, a portion of the shock front crossing the interface accelerates to velocity V_2 in a steplike manner, and after this its velocity remains constant. Due to the acceleration, this portion advances faster compared to the

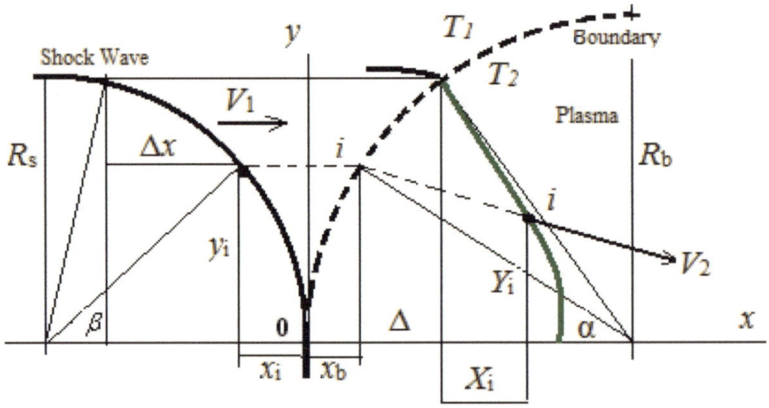

Figure 1.
Shock-plasma interaction diagram in the vertical plane of symmetry. As the initially spherical shock progresses through the spherical interface (dashed curve), its shape gradually deforms (green curve). Only the upper part of the diagram is shown.

reminder that is still propagating in the colder media thus resulting in the continuously increasing front stretching toward the hotter medium [18].

The shock front development proceeds in two stages, first being affected by the conditions on the interface and second in the plasma volume. The shock refraction resulting in an increase of the absolute value of the shock velocity along with its vector rotation (at refraction angle γ) occurs at the moment when the shock front crosses the plasma interface. As the refracted shock continues to propagate in hotter medium, its dynamics is determined by the parameter distribution in the plasma volume [17, 19, 33, 35]. Even though the changes in the shock structure become visible only during this time, they are the still consequences of the interaction at both stages: the conditions on the interface are necessary to trigger the front instability, and the gas volume effects provide the means necessary for its positive dynamics.

The relationship between the incident (x_i, y_i) and refracted (X_i, Y_i) shock front coordinates at a point of the interaction i has been derived in [33]. To recast it in a dimensionless form, the coordinates can be scaled with the plasma sphere radius R_b, the shock velocity with V_1, gas temperature with T_1, Mach number with M_{1n}, and time t with the characteristic time $\tau = R_b/V_1$:

$$\overline{X_i} = (\overline{v}\cos\gamma - 1)(n - (\overline{x_i} + \overline{x_b})) + \overline{\Delta x}, \overline{Y_i} = \overline{y_i} - \overline{v}\sin\gamma \bullet (n - (\overline{x_i} + \overline{x_b})) \qquad (1)$$

Here $n = t/\tau$ is the dimensionless time, $0 < n < 2$, α is the local incidence angle at the interaction point (**Figure 1**), $\overline{x_b} = 1 - \cos\alpha$, $\overline{\Delta x} = (R_s/R_b)(\cos\beta - \eta)$, the parameter

$$\eta = \frac{2(R_s + R_b)(R_s - nR_b) + n^2 R_b^2}{2R_s[(R_s + R_b) - nR_b]} \qquad (2)$$

and the bar over the variable means its dimensionless equivalent. The dimensionless shock velocity

$$\overline{v} = V_2/V_1 = \sqrt{\overline{T}\left(\overline{M}\right)^2 \cos^2\alpha + \sin^2\alpha} \qquad (3)$$

and the refraction angle $\gamma = \alpha - \tan^{-1}\left(\sqrt{\overline{T}\overline{M}}\tan\alpha\right)$ are determined by the problem geometry, heating intensity $\overline{T} = T_2/T_1$, and the ratio of normal components of Mach numbers in the two media $\overline{M} = M_{2n}/M_{1n}$ that account for the shock reflections off the interface.

The Mach number ratio for normal incidence can be obtained using the refraction equation from [36] that was derived assuming steplike temperature T_2/T_1 changes across the interface (a "sharp" interface)

$$\frac{1}{M_{1n}(k-1)}\left\{\left[2kM_{1n}^2 - (k-1)\right]\left[(k-1)M_{1n}^2 + 2\right]\right\}^{\frac{1}{2}} \times \left\{\left[\frac{2kM_{2n}^2 - (k-1)}{2kM_{1n}^2 - (k-1)}\right]^{\frac{k-1}{2k}} - 1\right\}$$
$$= M_{1n}\left(1 - \frac{1}{M_{1n}^2}\right) - M_{2n}\left(1 - \frac{1}{M_{2n}^2}\right)\left(\frac{T_2}{T_1}\right)^{\frac{1}{2}} \qquad (4)$$

and the normal and tangential components of the Mach numbers are related as $M_{1n} = M_1\cos\alpha$, $M_{1t} = M_1\sin\alpha$, and $M_{2t} = M_{1t}\sqrt{T_1/T_2}$. The "sharp" interface assumes back reflections off it and associated losses of the shock energy; thus, the ratio $M_2/M_1 < 1$. For an interface of a “smooth” type, when parameters across its thickness change slower than $1/x^2$, there are no losses associated with the shock reflection. Thus, M_2/M_1 approaches the unit, and the refraction effects become stronger [33]. An intermediate boundary type, when the parameters across its thickness can change faster than $1/x^2$ (an “extended” interface), has been considered in [34]. For a particular case of the exponential gradient in the parameter change, Eq. (3) can be replaced with more general $V_{2x}/V_1 = (T_2(x)/T_1)^\beta$, where $\beta = 1/2 - 1/\Sigma$ and $\Sigma = \left(\sum_{n=0}^{10} n\alpha_n + \sum_m m\alpha_m\right) = 53.58$ is the numerical constant calculated from the fit coefficients α_n and α_m in the equation:

$$\frac{T_2}{T_1} = \sum_{n=0}^{10} \alpha_n \left(\frac{M_1}{M_2}\right)^n + \sum_m \alpha_m \left(\frac{M_1}{M_2}\right)^m \qquad (5)$$

The numerical constants are valid within the limits for M_1 = (1.6–2.4), T_2/T_1 = (0–75), and k = 1.4 (air). The power coefficients m in the second sum run the values m = 1/16, 1/8, ¼, and ½. In Eq. (5), the coefficient β approaches exactly ½ for a limiting case of a smooth boundary and a value much smaller than ½ for another limit of a sharp boundary [34].

3. Basic modifications in the flow: numerical simulation results and comparison with experiment

To demonstrate how the shock refraction on the interface with plasma triggers the chain of consecutive transformations in the flow, the model was numerically run for three types of plasma density distribution that would cover the most common ways of the plasma production. A relatively uniform distribution law is commonly observed in microwave discharges [37] or when the plasma spot was initiated at a considerable distance from the body shock. The exponential density distribution can be established during the exothermal expansion, for example, in large-area plasma sources created with internal low-inductance antenna units [38], detonation, or the ultra-intense laser-induced breakdown in a gas [39]. And the power law density distribution can be found in the heated layers/spheres in thermodynamic equilibrium (for both media and radiation) in the presence of radiative heat conduction, as, for example, on stellar surfaces where it is formed as the result of the combined action of gravity, thermal pressure, and radiant heat conduction [24]. On the laboratory scale, the examples of such systems include: the gas of radiative spherical cloud when simultaneously considering the mechanical equilibrium and radiative transfer, in the diffusion approximation; a planar problem of the sudden expansion into vacuum of a gas layer with a finite mass and constant initial gas distribution; a problem of sudden expansion of a spherical gas cloud into vacuum; isentropic flows for which there is a class of self-similar solutions, such as for a strong explosion on a solid surface due to other body impact with generation of vapor cloud expanding into vacuum [40]; vaporization of the anode needle of a pulsed x-ray tube caused by a strong electron discharge [41]; explosion of wires by electric currents in vacuum systems; spark discharge in air in the early stages; the gas area near the edge of a cooling wave; motion of gas under the action of an impulsive load; the spherical shock wave implosion; a problem of bubbles collapsing in a liquid [40]; in the gas behind a weak blast wave [42].

The main difference in the treatment imposed by the nonhomogeneous plasma parameter distributions is that the refracted shock velocity becomes time dependent and the system of Eqs. (1)–(3) must be substantially modified. While the plasma parameter distribution will vary, the cold gas parameters will be always considered as distributed homogeneously. For easier comparison, all the simulation results presented below will be obtained for the same incident shock strength, heating intensity, and the interface parameters, with $M_1 = 1.9, T_1 = 293$ K, $T_1/T_2 = 0.10$, equal radii $R_s = R_b$, adiabatic index $k = 1.4$ (air), and smooth boundary type. The timely order in the sequence of the flow modification stages demonstrated below follows from the model logic and agrees well with that observed in experiments: the shock front distortion and its weakening; flow parameter redistribution and pressure drop in the post-shock flow, followed with the body drag reduction; vortex generation in the plasma volume; and finally deformation or collapse of the plasma bubble.

3.1 Shock front distortion

When the temperature/density in plasma is distributed uniformly, the relations Eqs. (1)–(3) apply. Results of numerical simulation in **Figure 2** obtained for this case demonstrate the shock front distortion as it progresses through the hot plasma sphere, at different propagation times starting at $n = 0.05$ (the most left curve) through $\Delta n = 0.05$ time intervals. To highlight the size of the interface effect, the results are plotted comparatively, with the upper part of the diagram corresponding to the smooth, lower, and sharp type of the interface.

The initially spherical front acquires a nearly conical shape stretched in the propagation direction, in good agreement with the experimental observations. The most central part of the front (near the longitudinal symmetry axis) is affected by significant stretching due to longer interaction times and smaller angle of incidence. The curvature sign changes from negative to positive, with the inflection point location tending the intermediate area off the axis. Comparison between upper and lower curves shows that both types of the boundary produce identically shaped fronts; however, the smooth boundary results in stronger effect.

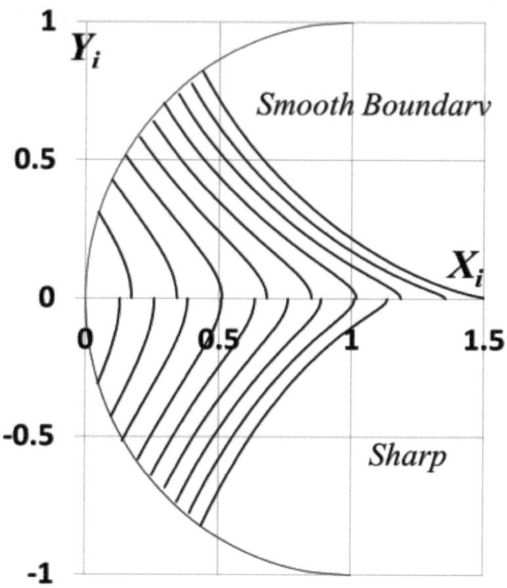

Figure 2.
The shock front modification in homogeneous plasma, at several interaction times. The shock is incident from left to right. The outside part of the shock remains spherical (not shown in the picture).

In the case of exponential plasma density distribution, the refracted shock velocity becomes dependent on time. Such a problem was considered in [19] assuming that the plasma density is exponentially decreasing in the shock propagation direction only, $\rho = \rho_{00}\, Exp\,(-x/z_0)$, starting from some finite value ρ_{00} at the leftmost point of the interface (z_0 is the characteristic length). The approximation neglecting the density change in the transverse direction can be applied, for example, to the case of plasma created by a laser sheet, with its body extended in the longitudinal direction. The task is reduced to obtain two separate solutions to the problem that must be tailored at the interface. The first solution models the shock wave refraction on the interface giving its speed and direction, and the second one describes its propagation in the inhomogeneous medium after crossing the interface. Using the approach [19] modified here for the sphere-to-sphere problem geometry, the time-dependent system of equations for the front's surface coordinates can be derived.

$$X_i = \sigma(t - t_{0i} + t_\lambda)^{2/5} + \varepsilon(t - t_{0i} + t_\lambda)^{4/5} - x_0,\ Y_i = y_i - \sigma(t - t_{0i} + t_\gamma)^{2/5} - \sigma t_\gamma^{2/5} \tag{6}$$

where $x_0 = \sigma t_\lambda^{2/5} + \varepsilon t_\lambda^{4/5}, t_{0i} = \frac{(x_i + x_b)}{V_1}, t_\gamma = \left(5V_1 \cos\alpha_i \sin\gamma_i \sqrt{\frac{T_2}{T_1}}\right)/$ $\{2\sigma\cos(\alpha_i - \gamma_i)\}$ and the parameter t_λ is the solution of the following equation

$$\frac{5\,V_1 \cos\alpha_i \cos\beta_i}{2\,\sigma\cos(\alpha_i - \gamma_i)}\sqrt{\frac{T_2}{T_1}} = t_\lambda^{-3/5} + \frac{2\varepsilon}{\sigma} t_\lambda^{-1/5} \tag{7}$$

The parameters σ and ε in Eqs. (6) and (7) are related to the effective explosion energy E in the thermal spot and the scale of the density gradient z_0 as $\sigma = \varsigma\,(E/\rho_{00})^{1/5}$ and $\varepsilon = (K/z_0)\sigma^2$, and the numerical parameters $\varsigma = 1.075$ and $K = 0.185$ are borrowed from Ref. [43].

The simulation results presented in **Figure 3a** were obtained for the interaction times starting at $n = 0.05$ through $\Delta n = 0.05$ time intervals, $z_0 = 0.225$ cm, $\rho_1/\rho_{00} = 10$, $R_b = 0.1$ cm, and the parameters $\alpha = 7$ and $\beta = 402.79$ correspond to the specific explosion energy $E/\rho_{00} = 11.707 \cdot 10^3$J. The most striking result in this case is that such a density profile can generate virtually perfect conical shock fronts, exactly as observed in the experiment [13].

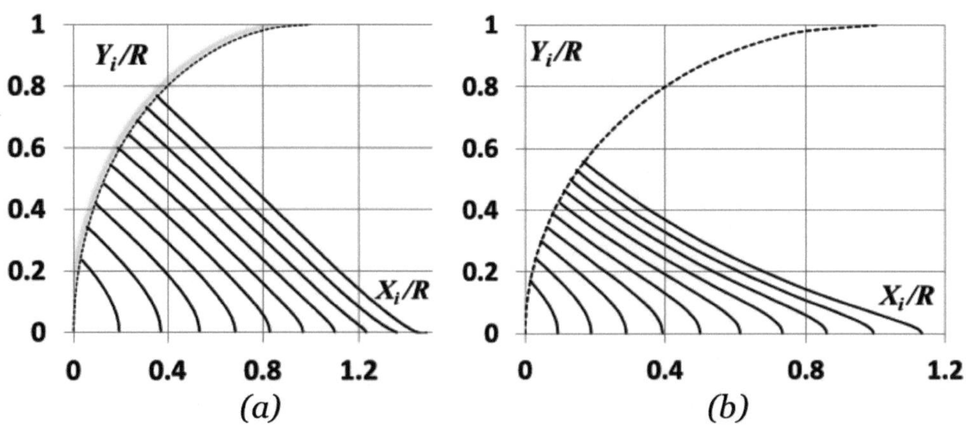

Figure 3.
Shock front deformation at several interaction times, for the cases of exponential (a) and power law (b) density distributions. Note twice as short interaction times in graph (b) compared to graph (a).

When a plane shock wave propagates through a gas with the density that drops to zero according to a power law $\rho \sim x^N$, the so-called energy cumulation effect takes place [44]. In the gas-dynamical approximation, a strong plane shock wave propagating in such a medium accelerates very quickly accumulating virtually infinite energy. This interesting phenomenon, as applied to the shock refraction problem, was considered in [35] assuming the plasma density as changing in the longitudinal (x-) direction only. Applying it to the present geometry, the refracted shock coordinates can be determined as

$$X_{2i} = a_i - (G_i/b)(t_{0i} - t)^b,\ Y_{2i} = y_i - (V_2/V_1)\sin\gamma[nR - (x_i + x_b)] \tag{8}$$

the shock velocity component $V_{2xi} = G_i(t_{0i} - t)^{b-1}$, $a_i = a - R_b(1 - \cos\alpha_i)$,

$$G_i = \frac{V_1 \cos\gamma_i \sqrt{\cos^2\alpha_i \cdot (T_2/T_1)(M_{2n}/M_{1n})^2 + \sin^2\alpha}}{[t_{0i} - ((x_i + x_b)/V_1)]^{b-1}} \tag{9}$$

$$t_{0i} = \frac{b(a - R_b(1 - \cos\alpha_i))}{V_1 \cos\gamma_i \sqrt{\cos^2\alpha_i \cdot (T_2/T_1)(M_{2n}/M_{1n})^2 + \sin^2\alpha}} + \frac{(x_i + x_b)}{V_1} \tag{10}$$

and the shock velocity ratio V_2/V_1 is determined from Eq. (3). Here a and a_i are the maximum (from the point O in **Figure 1**) and the local (from the boundary, at the periphery from the axis) distances to the zero density plane correspondingly, t_{0i} is the local time of the shock front portion travel to the boundary, the constant b = 0.59 was determined in [44], and the constant N = 3.25 is taken from [40].

The system of Eqs. (8)–(10) was run for the distance to the zero density plane a = 3.0R_b, at times starting at n = 0.025 through the equal increments Δn = 0.025 (note that the interaction times here are twice as short than in **Figure 3a**). The results presented in **Figure 3b** demonstrate considerably stronger stretching of the shock front per unit of time due to significant front acceleration in the plasma area.

Thus, as seen from **Figures 2** and **3**, the most common type of the front deformation for all three types of the density distribution is its continuous stretching. To make more exact conclusion about the front distortion, the local front inclination angle was computed, for M_1 = 1.9, T_2/T_1 = 10.0, and the smooth type of the interface.

In **Figure 4a**, the angles are presented for the uniform law of distribution, at times starting at n = 0.05 through the intervals Δn = 0.05, and the radii of the incident shock front and the interface are R_s = R_b = 0.3 cm. The curves in **Figure 4b** are for the angles corresponding to the exponential law of distribution, at the same

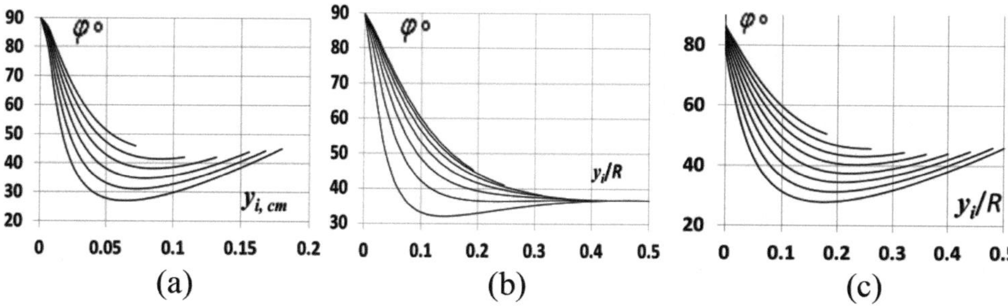

Figure 4.
The shock front inclination angle φ vs. distance y_i, at several interaction times, for the uniform (a), exponential (b), and power law (c) density distributions. The time sequence for the curves is from upper to lower.

times as in **Figure 4a**, for z_0 = 0.225 cm and σ = 7, and that in **Figure 4c** are for the power law of the distribution at shorter times, n = 0.025 through the intervals Δn = 0.025. The time sequence for the curves in all graphs is from upper to lower. The minimum angle corresponding to a location next to the longitudinal symmetry axis is common for all three types of the distribution. The angle increase to 90° as $y_i > 0$ is due to normal incidence at this point.

3.2 Pressure drop and drag reduction

In experiments, at the time when the gas pressure around the body significantly drops, recordings also show that the image of the front part of the shock wave becomes blurred or invisible and the deflection signals appear weaker and widened with time (shock "dispersion" and "disappearance") [10, 12]. As was seen above, continuous stretching of the shock front during its advancement through the plasma transforms it from being nearly normal into an inclined one. Due to the inclination, its progressively weakened intensity results in vanishing compression across the shock. As a result, the gas pressure in the flow in front of the body is lowered. If this continues, the shock wave degenerates in an ordinary pressure wave that becomes unidentifiable in the experiments.

The effect of the shock deformation on the pressure drop can be estimated through the change in the Mach number. For the homogeneous density distribution in plasma, the local inclination angle of the front φ, as defined through the tangent line, can be used to estimate the local parameter jump across the front. For this, the expressions for gas parameter variation across the normal shock front can be corrected, with the total Mach number replaced with its normal to the front component $M_{2n} = M_2 \sin(\varphi - \gamma)$. Then the pressure jump across the refracted shock $p_{21}^{(ref)}$ normalized to the one across the incident shock $p_{21}^{(inc)}$:

$$p_{21} = \frac{p_{21}^{(ref)}}{p_{21}^{(inc)}} = \frac{(2k/(k+1))M_{2n}^2 - (k-1)/(k+1)}{(2k/(k+1))M_1^2 - (k-1)/(k+1)} \tag{11}$$

In defining this parameter, the incident shock was taken as normal. This approximation works best for the area of the most strong interaction (near the symmetry axis area) or when the thermal spot dimensions (R_b) are considerably smaller compared to the incident shock radius R_s.

In **Figure 5a**, the simulation results for pressure p_{21} in the uniform density distribution case is presented, at times starting at n = 0.05 through Δn = 0.05 increments. Matching with the data in **Figures 2a** and **4a**, it is seen that the pressure

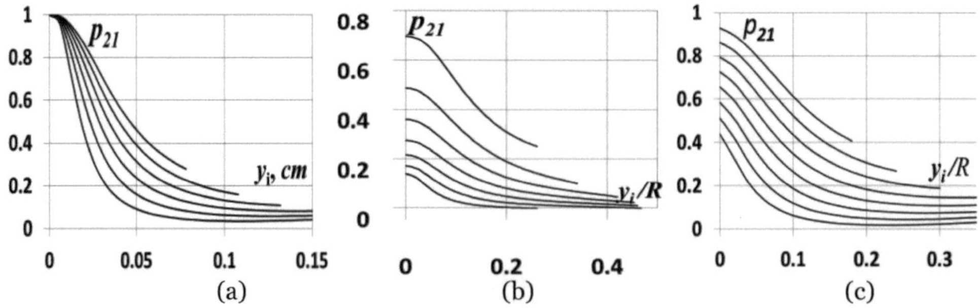

Figure 5.
Pressure ratio p_{21} vs. coordinate y_i, for the following plasma density distribution laws: (a) uniform, (b) exponential, and (c) power law. The time sequence for the curves is from upper to lower.

distribution quite closely follows the inclination angle change suggesting that this angle is the key parameter in the phenomenon.

On the axis ($y_i = 0$), the pressure ratio stays the same as for the incident shock due to normal position of the front relative to the propagation direction ($\varphi = 90°$).

For the exponentially distributed density, the Mach number

$$M_{2x}(t) = \left[\sigma(t - t_{0i} + t_\lambda)^{-3/5} + \varepsilon(t - t_{0i} + t_\lambda)^{-1/5}\right] / \left(20.043\sqrt{10T_1 \exp\{X_{2i}(t)/z_0\}}\right) \tag{12}$$

becomes time dependent, and the front deformation develops with an acceleration. The results for p_{21}, obtained for times starting at $n = 0.05$ through $\Delta n = 0.05$ increments, $z_0 = 0.225$ cm, $\sigma = 7$, and $R_s = 0.1$ cm, are presented in **Figure 5b**. For the most part of the shock front, the pressure significantly drops, with the distribution closely matching the inclination angle change (**Figure 4b**). The pressure ratio levels are in good agreement with the observations of cone-shaped fronts [13] where the reduction by three to four times in the pressure in the central area of the shock front and almost zero pressure at the stagnation point is typical.

Compared to the two previous cases, the effect of the power law density distribution is stronger (**Figure 5c**): the front becomes deformed to a larger degree and moves considerably faster (note twice as short interaction times, starting at $n = 0.025$, with $\Delta n = 0.025$ increments). The pressure ratio drops to the levels between 0.25 and 0, and the compression in the shock becomes weaker as it approaches the 0 density point. This is possible despite the sharp increase in the velocity because the density ahead of the wave decreases faster, similarly to the case of a plane shock propagating through an unbounded plasma medium [40].

Since the gas pressure in front of the moving body becomes significantly lowered, the decrease in the drag follows. In the spherically symmetrical case, the drag can be numerically estimated by summing the longitudinal pressure component over the body's surface $d = \sum_i 2\pi r_i p_i \cos\alpha_i - d_b$, where α_i is the local angle of incidence on the body and d_b is the summing result for the back body surface. While the total drag experienced by the body is dependent on its shape, it is still linearly proportional to the ambient pressure and will decrease accordingly. The drag reduction about 2–2.5 times recorded in experiments, at the times when the shock modification occurred, confirms these conclusions.

3.3 The parameter redistribution in the flow and generation of vorticity

As a result of refraction, the initially horizontal shock velocity vector rotates at local refraction angle γ_i, and thus the velocity acquires an additional, location-dependent y-component. Then the velocity components for the flow behind the shock

$$v_x = v_n \sin\varphi + V_2 \cos(\varphi - \gamma)\cos\varphi,\ v_y = v_n \cos\varphi - V_2 \cos(\varphi - \gamma)\sin\varphi \tag{13}$$

result in non-zero vorticity [45] that can be calculated at specific locations along the shock front surface as

$$\omega_i = \omega_i(X_i, Y_i) = \left(\partial v_x/\partial y - \partial v_y/\partial x\right)_{x=X_i,\, y=Y_i} \tag{14}$$

Thus, in addition to the rotation of the shock velocity vector at an angle γ, there is a turn for the flow velocity relative to its initial propagation direction, at the angle

$g = tan^{-1}(v_y/v_x)$. The shock wave and the flow behind it will diverge at a relative angle $\xi = tan^{-1}(v_y/v_x) - \gamma$ (double refraction) giving rise to a non-zero circulation in the hot gas volume.

Numerical results for dimensionless vorticity $\overline{\omega} = \omega_i/(V_1/R_b)$ developing in the flow with the homogeneous plasma parameter distribution vs. the coordinate Y_i are presented in **Figure 6a**, for times between $n = 0.18$ and 0.42 with $\Delta n = 0.03$ increments, and the last curve corresponds to $n = 0.44$ chosen for better resolution on the graph. The time sequence for the curves is in the up-direction. Only the upper half of the picture is shown; thus, the results on the graph correspond to a vortex sheet with rotations in both halves of the picture in opposite directions (in cylindrical geometry) or a toroidal vortex ring if considered in the spherical geometry.

Due to a small size of the structure, the Y-coordinate was scaled with the factor of 10^2, and thus the picture is greatly zoomed in on the narrow region next to the symmetry axis. The spike in the vorticity intensity observed in close proximity to the symmetry axis exactly corresponds to the location where abrupt change in the shock front structure occurs (**Figures 2a** and **4a**). This trend is pointing at the fact that the local rate of change of the front inclination angle φ is the key factor in the origin of the vorticity. The nonlinear growth of the vortex intensity (**Figure 6**) demonstrates the possibility of a strong positive dynamics in its development.

The specific effect of the exponential density distribution has been studied for the interaction times between $n = 0.05$ and 0.40 through $\Delta n = 0.05$ increments, and the last two curves correspond to $n = 0.43$ and 0.45. (**Figure 6b**). Due to practically no areas on the shock front where its shape would change sharply enough (**Figures 2b** and **4b**), except a very narrow region near the symmetry axis, significant vorticity develops mostly in this region corresponding to the very tips of the fronts. The maximum vorticity intensity is about two orders of magnitude less compared to the levels found in the uniform density distribution case. Considerably weaker vorticity and a very small size of the structure found here can probably explain why sometimes the vortex system "does not develop" in experiments even though noticeable changes in the shock structure and pressure/drag reduction are present.

In the presence of the energy accumulation effect (power law density distribution), with more stretched fronts and softer distribution of the front inclination angle φ, the vorticity develops slower (**Figure 6c**), but still, its intensity steadily grows with time at all locations on the front. The curves on the graph correspond to times $n = 0.025$–0.250 with 0.025 increments and the distance $a = 3.0R_b$. The vorticity originates further from the symmetry axis confirming a larger size of the vortex structure. As in other density distribution cases, the maxima of vorticity (centers) shift toward each other as the shock advances through the volume, following the same trend for the sharpest bending on the front and the shock and flow rotation angles to move closer to the axis [45]. The total vorticity intensity

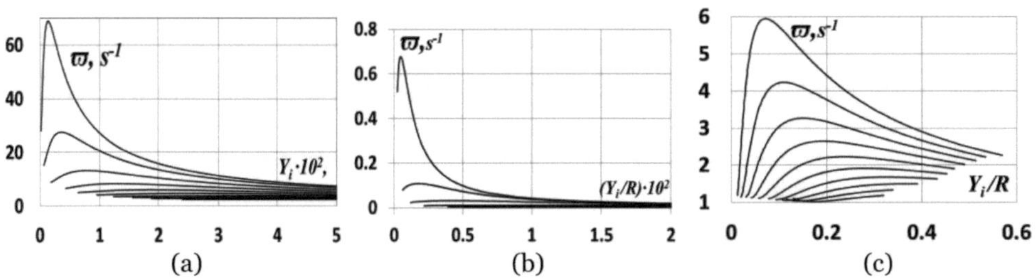

Figure 6.
Dimensionless vorticity $\overline{\omega}$ vs. Y_i, for the uniform (a), exponential (b), and power law (c) density distributions. The time sequence is from lower curve to upper.

integrated over the whole plasma volume becomes considerably higher compared to the uniform density case because of the contribution from the regions located further from the axis. The same strong correlation between shock deformation, vortices of similar size, rotational direction, and topology with pressure/wave drag drop evolving in the same sequence and within the same time frames was confirmed in [46].

3.4 The interface stability problem

The existence of phenomenological connection between the shock's and the interface stability makes the chain of the transformations in the flow to continue. Since the flow instabilities take place upstream from the interface, the perturbations to the flow parameters propagating downstream, toward the interface, will disturb it. The overall pressure drop behind the refracted shock continuously mounting with time is responsible for the sucking effect resulting in the large-scale interface perturbation, moving it closer to the shock. The positive and essentially nonlinear dynamic in the pressure perturbation evolution [17] will support amplification of this global perturbation to the interface and thus determines the pattern in the interface instability structure. With increasing distortion of the interface, Kelvin-Helmholtz (KH) shearing instability may start to contribute resulting in the characteristic mushroom shapes of the interface perturbations [47]. The initial instability pattern associated with the global pressure drop behind the refracted shock will be of a larger scale, and the KH instability turning on at later stages would finally determine the smaller characteristic structure.

4. Conclusion and controlling the drag

The sequence of interconnected stages of the flow modification described here, in fact, is a complex way of the shock flow instability development that is triggered via the single mechanism of the shock refraction on an initially disturbed (curved) interface. At the moment of exiting the interface disturbance, the shock wave reaches a new stable state characterized by a new front structure and the gas parameter distribution behind it. The instability starts to develop in the form of a wavelike shock front stretching into the lower density plasma, and the global pressure drop is the consequence of the weakened shock wave. Pressure perturbations caused by the shock stretching result in the loss of stability of the flow behind it that eventually organizes into an intense clockwise rotating vortex structure. If the plasma density is nonuniform, the transition in the form of front stretching exhibits the pattern of motion that prevails the principle of exchange of stabilities, so the instability sets in as a secondary flow. Regardless of the interaction geometry, the instability mode is aperiodical and unconditional, and thus either a transition to another stable state or continuous development as a secondary flow is possible. The marginal state condition $\{T_2/T_1 > (M_1/M_2)^2, |\chi| > 0\}$ is the only requirement to trigger the instability [17], where the minimum heating requirement accounts for losses due to shock reflections off the interface. Independence of the instability locus on the plasma density distribution identifies the interface conditions as the sole triggering factor, though the density gradient can discriminate between qualitatively different outcomes at later stages of the instability evolution. With T_{21} fixed, in the uniform density case, the perturbation growth rates are determined solely by the interface perturbation curvature χ that, for some geometries, can be replaced with the *relative* curvature between the shock front and the interface. The specific wave nature of the instability dissipation, when the overstretched shock perturbation decays via the degeneration into an acoustic wave, allows the shock instability to decay even though the viscous damping mechanisms are not available in the flow (except the shock width layer). The initial state shock wave energy, in this case, converts into rotational energy of the flow thus continuously supporting the developing vortex structure until the perturbed shock vanishes.

The advantage of the model presented here is that it provides useful insight on the phenomenological origin of the drag reduction and its connection to the series of other phenomena that accompany the interaction. Its simple analytical solutions can be used for quick estimates and thus avoid the complexity of more exact numerical calculations. The dimensionless form of the equations and similarity law found in the spatial and temporal evolution of the flow perturbations with respect to the interface curvature [17] allow conclusions to be applied to any scale of the interaction. Though the model here was specifically developed for the spherical or sphere-to-flat (as a limiting case) geometry of the interaction, the equations can be readily modified for other configurations. The shock reflections off the interface can be accounted in the model to its various degrees with the Mach number ratio M_{2n}/M_{1n} different from the unit [34], thus accommodating various types of the interface. Note that even plasma as a medium inside the heated area was considered, the nature of the effect is purely thermal and hence the model can be applied to conditions where only thermal heating in a neutral gas is present.

It was shown here that gas parameter redistribution in the flow behind the refracted shock front, including the global pressure drop, is the cause for the vorticity generation rather than its consequence. Thus, the pressure drop and the vortical perturbations in the post-shock flow can be considered rather competing because of the locally developing secondary flows resulting in an effective mixing. This eventually equalizes pressure in different parts of the volume and thus quenches (at least partially) the pressure drop and large-scale sucking effect on the interface. Thus, if pressure lowering is thought, the vortex generation must be minimized. Then the parameters of the interaction must be chosen in such a way that the shock front would be stretched considerably and as possible evenly, without sharp bends causing intensive vorticity. Spherical geometry and the exponential type of the plasma parameter distribution would be the best fit for this purpose as it was shown to generate vorticity of either micro-size and/or micro-intensity. Thus, the choice of the plasma parameter distribution alone can offer the advantage of depositing less energy of a smaller volume and closer to the body.

Other applications where the vorticity development must be avoided include magnetized target fusion experiments, for example, where, during the implosion of an inertial confinement fusion target, the hot shell materials surrounding the cold D-T fuel layer are shock-accelerated [25]. Mixing of the shell material and fuel is not desired in this case, and efforts should be made to minimize any abrupt imperfections or irregularities on the shock front. Considering a sharp type of the interface can be helpful in such cases since the refraction effect can be up to 40% weaker [18].

Contrary, in the applications where vorticity is beneficial for better mixing in the flow, sharp distortions spanning a larger portion of the shock front are necessary, and the uniform and the power law density distributions could be more desirable in this case. The examples include combustion where the burning speed is controlled by introducing turbulence in the flow [22]. A supersonic combustion in a scramjet may also benefit from vorticity in the flow as the fuel-oxidant interface is enhanced by the breakup of the fuel into finer droplets. Studies of deflagration to detonation transition (DDT) processes show that introduction of vorticity can result in an increase in burning velocity and detonation. In this case, introducing small perturbations of the interface that will cause sharp distortions on the shock front can be suggested [17].

In optimizing the ratio R_s/R_b, it should be noted that the interaction can be less significant if the size of the thermal spot becomes small compared to the shock front dimensions. It is not only because of a relatively smaller plasma volume or deposited energy but also due to smaller relative curvature of the shock front as seen from the plasma boundary (approaching a nearly flat shape as $R_s/R_b \rightarrow 0$). This results in shorter interaction time and consequently weaker effect. Higher temperature/density step across the interface and the smooth interface type can be also used to strengthen the effect.

In discussing validity of the model for a particular application, the following factors must be taken into consideration. First, the model assumes that the plasma region is not expanding, i.e., that the gas is in an equilibrium state. With the interaction times on the order of 0.1 μs and in the temperature range typical for many thermal energy deposition experiments, the approximation works well, for example, for diatomic gases whose vibrational relaxation times are on the order of tens of microseconds or longer (at atmospheric pressure). However, for considerably elevated gas temperatures or pressures and in specific gases, the relaxation times may become comparable to the interaction time, and the model would need a correction for non-equilibrium effects.

Second, the model does not assume various degree deformation of the plasma region during the interaction time and thus should be considered as a valid approximation only when the coupling between the shock flow and the interface instability can be neglected. The model also assumes the ideal gas conditions. For the pressures typically developed in the shocks, the gas state equation can be still valid. However, the parameter jump relations across the shock in real gases may depart from their ideal equivalents significantly, and Eq. (4) must be corrected. An analysis of such effects must be done, and the criteria of ideality have to be applied and tested in the experiments used for comparison. For most energy deposition conditions though, the approach developed here appears to be quite adequate as, within the time frame of the interaction, the model predictions in the form of consistent time and dimension relationship are clearly visible in most of the experiments [12, 22].

The shock refraction on an interface was considered here as the main mechanism of the wave drag reduction. The contribution of other thermal or nonthermal secondary mechanisms should be accounted for as well. In addition to those mentioned in the introduction, these can include the mechanisms participating in the formation of the thermal equilibrium and definite parameter distribution in the plasma during rapid expansion of the thermal spot. A blast wave produced at the early stages of the plasma creation can result in the pressure increase on the body at the first moment followed by the main pressure and the drag drop [48]. This may interfere with the shock refraction effects and overlap with the processes described in the model. The mechanisms controlling the effectiveness of the energy transfer to the gas during its deposition to the flow should be taken into account for better optimization.

The results of this work can be found useful in a number of other major applications. With optimized energy deposition, decreased pressure with an extended rise time on the ground can be used to alleviate the sonic boom or reduce the heating of a space craft at atmospheric entry. Other potential areas of application include magnetohydrodynamics (MHD), plasma chemical reactors, and molecular lasers for shock wave structure control; rocket plume optimization; shock wave-assisted combustion, where a shock wave can be used to affect the ignition conditions in the gas; plasma-based piloting and combustion enhancement that can be used toward the design of efficient combustors for hypersonic propulsion systems, combustors for gas turbine, diesel engines, environmentally clean combustors, and spark inhibition; localized flow problems, such as the Edney type IV shock interaction and the control of shock structures formed in the high-speed engine inlet; and astrophysics for understanding the dynamics of shock waves generated in the stellar interior.

Author details

Anna Markhotok
Physics Department, Old Dominion University, Norfolk, VA, USA

*Address all correspondence to: amarhotk@phys.washington.edu

References

[1] Bushnell D. Shock wave drag reduction. Annual Review of Fluid Mechanics. 2004;**36**:81-96. DOI: 10.1146/annurev.fluid.36.050802.122110

[2] Carlson HW, Mack RJ. Studies of leading edge thrust phenomena. In: 18th AIAA Aerospace Sciences Meeting; Pasadena, CA: AIAA Paper; 1980. pp. 80-0325.

[3] Jones RT. Aerodynamic design for supersonic speeds. In: Karman V et al., editors. Advances in Aeronautical Sciences. New York: Pergamon; 1959. pp. 34-51

[4] Jones RT. The flying wing supersonic transport. Aeronautical Journal. March 1991;**95**(943):103-106

[5] Bushnell DM. Supersonic aircraft drag reduction. In: AIAA 21st Fluid Dynamics and Plasmadynamics Lasers Conference; Seattle, WA: AIAA Paper; 1990. pp. 90-156

[6] Kuflan RM. Application of favorable aerodynamic interference to supersonic airplane design. SAE Technical Paper; 901988; Warrendale, PA: SAE; October 1990

[7] Bur R, Corbel B, Delery J. Study of passive control in a transonic shock wave/boundary layer interaction. AIAA Journal. 1997;**36**(3):394-400

[8] Reneaux J, Coustols E, Wave drug reduction technologies. In: Onera-DLR Aerospace Symposium; Paris; 1999

[9] Austin F, Siclari MJ, Van Nostrand WC, Weisensel GN, Kottamasu V, Guiseppe V. Comparison of smart wing concepts for transonic cruise drag reduction. In: Proceedings of SPIE-The International Society for Optical Engineering; 3044; Bellingham, WA; 1997. pp. 33-40

[10] Kuo SP. Shock wave modification by a plasma spike: Experiment and theory. Physica Scripta. 2005;**71**:535-539

[11] Zukovski E, Spaid F. Secondary injection of gases into a supersonic flow. AIAA Journal. 1964;**2**(10):1689-1696

[12] Erdem E, Kontis K, Yang L. Steady energy deposition at Mach 5 for drag reduction. Shock Waves. 2013;**23**(4): 285-298. DOI: 10.1007/s00193-012-0405-8

[13] Satheesh K, Jagadeesh G. Experimental investigations on the effect of energy deposition in hypersonic blunt body flow field. Shock Waves. 2008;**18**:53-70

[14] Kremeyer K, Sebastian KA, Shu C-W. Computational study of shock mitigation and drag reduction by pulsed energy lines. AIAA Journal. 2006;**44**: 1720-1731

[15] Guvernyuk S, Samoilov A. Control of supersonic flow around bodies by means of a pulsed heat source. Technical Physics Letters. 1997;**23**:333-336

[16] Schulein E, Zheltovodov A, Pimonov E, Loginov MS. Experimental and numerical investigations of electric-arc air spikes for blunt and sharp bodies at Mach 5. In: International Conference on Methods of Aerophysical Research, ICMAR; 2008

[17] Markhotok A. A shock wave instability induced on a periodically disturbed interface. IEEE Transactions on Plasma Science. 2018;**46**(8): 2821-2830. DOI: 10.1109/TPS.2018.2848597

[18] Markhotok A, Popovic S. Refractive phenomena in the shock wave dispersion with variable gradients. Journal of Applied Physics. 2010;**107**: 123302

[19] Markhotok A, Popovic S. Redirection of the spherical expanding shock wave on the interface with plasma. Physics of Plasmas. 2014;**21**: 022105

[20] Marconi F. An investigation of tailored upstream heating for sonic boom and drag reduction. AIAA Journal. 1998;**333**:1-6

[21] Sasoh A, Ohtani T, Mori K. Pressure effect in shock-wave-plasma interaction induced by a focused laser pulse. Physical Review Letters. 2006;**97**: 205004

[22] Thomas GO, Bambrey RJ, Brown CJ. Experimental observations of turbulent combustion and transition to detonation following shock-flame interaction. Combustion Theory and Modelling. 2001;**5**:573-594

[23] Georgievskii PY. Transition to Irregular regimes of supersonic flows over bodies initiated by energy deposition. In: 43rd AIAA Aerospace Sciences Meeting and Exhibit; Reno, NV; 2005

[24] Frank-Kamenetskii DA. Non-adiabatic pulsations in stars. Doklady Akademii Nauk SSSR. 1951;**80**:185 (In Russian)

[25] Suponitsky V, Barsky S, Froese A. On the collapse of a gas cavity by an imploding molted lead shell and Ritchmer-Meshkov instability. Bonneville Place, Burnaby, Canada: General Fusion Inc.; 2013

[26] Ju Y, Shimano A, Inoue O. Vorticity generation and flame distortion induced by shock flame interaction. Symposium (International) on Combustion. 1998; **27**(1):735-741

[27] Bray KNC. In: Libby PA, Williams FA, editors. Turbulent Reacting Flows, Topics in Applied Physics. Vol. 44. Berlin: Springer; 1980

[28] Shneider M. Energy addition into hypersonic flow for drag reduction and steering. In: Atmospheric Pressure Weakly Ionized Plasmas for Energy Technologies, Flow Control and Materials Processing. August 22–24, 2011; Princeton, NJ

[29] Khorunzehenko V, Roupassov D, Starikovskii A. Hypersonic flow and shock wave structure control by low temperature nonequilibrium plasma of gas discharge. In: 38th AIAA/ASME/SAE/ASEE Joint Propulsion Conference and Exhibition; AIAA 2002-3569. 2002

[30] Kuo SP. Shock wave mitigation by air plasma deflector. Advances in aerospace science and technology. 2018; **3**:71-88

[31] Satheesh K, Jagadeesh G. Effect of concentrated energy deposition on the aerodynamic drag of a blunt body in hypersonic flow. Physics of Fluids. 2007;**19**(3):031701

[32] Batley GA, McIntosh AC, Brindley J. The baroclinic effect in combustion. Mathematical and Computer Modelling. 1996;**24**(8):165-176

[33] Markhotok A. The mechanism of wave drag reduction in the energy deposition experiments. Physics of Plasmas. 2015;**22**:063512

[34] Markhotok A, Popovic S. Shock wave refraction enhancing conditions on an extended interface. Physics of Plasmas. 2013;**20**:4

[35] Markhotok A. The cumulative energy effect for improved ignition timing. Physics of Plasmas. 2015;**22**:043506

[36] Alexandrov AF, Vidiakin NG, Lakutin VA, Skvortsov MG, Timofeev IB, Chernikov VA. Potential mechanism of shock-wave interactions with decomposing plasma of laser spark in the air. Zhurnal Tekhnicheskoi Fiziki. 1986;**56**:771

[37] Kessaratikoon P, Markhotok A, Brooke G, Popović S, Vušković L. Shock wave dispersions in microwave discharges. Bulletin of the American Physical Society. 2002;**47**:43

[38] Deguchi H, Yoneda H, Kato K, Kubota K, Hyashi T, Ogata K, et al. Size and configurations in large-area RF plasma production with internal low-inductance antenna units. Japanese Journal of Applied Physics. 2006;**45**: 8042-8045

[39] Bret A, Deutsch C. Beam-plasma electromagnetic instabilities in a smooth density gradient: Application to the fast ignition scenario. Physics of Plasmas. 2005;**12**:102702

[40] Zeldovich YB, Raizer YP. Physics of Shock Waves and High-Temperature Hydrodynamic Phenomena. New York and London: Academic Press; 1967

[41] Tsukerman V, Manakova M. Sources of short X-ray pulses for investigating fast processes. Soviet Physics Technical Physics (English Translation). 1957;**2**:353-363

[42] Bah G, Lee J. An analytical solution for blast waves. AIAA Journal. 1970;**8**: 271-275

[43] Andriankin EI, Kogan AM, Kompaneets AS, Krainov VP. Propagation of a strong explosion in an inhomogeneous atmosphere. Zhurnal Prikladnoi Mekhaniki I Tekhnicheskoi Fiziki. 1962;**6**:3-7 (In Russian)

[44] Sakurai A. On the problem of a shock wave arriving at the edge of gas, commune. Pure and Applied Mathematics. 1960;**13**:353-370

[45] Markhotok A. A mechanism of vorticity generation in a supersonic flow behind an interface. Fluid Dynamics Research. 2018;**50**(4):1-22. DOI: 10.1088/1873-7005/aac118

[46] Lashkov VA, Mashek IC, Anisimov YI, Ivanov VI, Kolesnichenko YF, Ryvkin MI, et al. Gas dynamic effect of microwave discharge on supersonic cone-shaped bodies. In: 42nd AIAA Aerospace Sciences Meeting and Exhibit; Reno, NV; AIAA 2004-0671. 2004

[47] Bruillette M. The Richtmeyer-Meshkov instability. Annual Review of Fluid Mechanics. 2002;**34**:445-468

[48] Knight D. Survey of aerodynamic drag reduction at high speed by energy deposition. Journal of Propulsion and Power. 2008;**24**(6):1153-1167

Chapter 6

Boundary Layer Theory: New Analytical Approximations with Error and Lambert Functions for Flat Plate without/with Suction

Chedhli Hafien, Adnen Bourehla and Mounir Bouzaiane

Abstract

In this work, we investigated the problem of the boundary layer suction on a flat plate with null incidence and without pressure gradient. There is an analytical resolution using the Bianchini approximate integral method. This approximation has been achieved by Lambert or Error functions for boundary layer profiles with uniform suction, even in the case without suction. Based on these new laws, we brought out analytical expressions of several boundary layer features. This gives a necessary data to suction effect modeling for boundary layer control. To recommend our theoretical results, we numerically studied the boundary layer suction on a porous flat plate equipped with trailing edge flap deflected to 40°. We showed that this flap moves the stagnation point on the upper surface, resulting to avoid the formation of the laminar bulb of separation. Good agreement was obtained between the new analytical laws, the numerical results (CFD Fluent), and the literature results.

Keywords: boundary layer suction, analytical approximation, Error function, Lambert function, trailing edge flap

1. Introduction

The advantage of the parietal suction is to delay the transition of the boundary layer to turbulence, reduce drag, and increase lift (avoid stalling) [1]. A method used in aviation is the multi-perforation of the walls. This type of boundary layer control is at the project stage on the wings of airplanes, such as the F-16XL and the A320, and starts to emerge in other industries. Equally, this technic is of interest in different engineering branches as the extraction of geothermal energy, nuclear reactor or electronics cooling system, filtration process, lubrication of ceramic machine parts, etc.

The theoretical and numerical study of laminar boundary layer over a flat plate without and/or with uniform suction is well introduced in the literature [2–8]. The modeling of the effect of this control technics can require analytical expressions of characteristic parameters of flow. Equally, theoretical investigations help understand the underlying physics of boundary layer suction and help to predict, with a minimum waste of time, its effect.

 IntechOpen

For the homogeneous suction, Schlichting and Bussmann [9] assumed that the longitudinal velocity gradient is null. Based on this hypothesis, he found an asymptotic solution which is expressed by the exponential function, but this profile is not valid in the region near the leading edge. To improve this solution, Preston [10] considered a family of parameters of velocity profiles having the Blasius profile and the asymptotic profile as a limit form. The solution obtained is more accurate by comparing it with the exact numerical solutions of Iglisch [11]. Palekar and Sarma [2] who applied the Bianchini approximate integral method [12] determined an analytical solution of the boundary layer profile expressed by the Error function in the case of nonuniform suction. This law was not compared with the real profile of Blasius. In the case of the uniform suction, the solution was obtained in an asymptotic form.

Kay [13] took velocity measurements out of a blower up a flat plate with uniform suction. The vertical velocity distribution was described by the exponential function. Aydin and Kaya [4] have considered finite difference approximations to resolve the boundary layer equations. Fang et al. [5] studied a similarity equation of the momentum boundary layer for a moving flat plate in a stationary fluid with mass suction at the wall surface. They provided a new solution branch for the Blasius equation. Recently, researchers have studied convergent and closed analytical solution of the Blasius Equation [14–16]. Wedin et al. [17, 18] have studied the effect of plate permeability on nonlinear stability of the asymptotic suction boundary layer. Zheng et al. [8] have proposed a solution of the Blasius equation expressed by two power series. They showed that the method for finding the closed analytical solution of Blasius equation was used in the regulation of the boundary layer injection and slip velocity.

The study of the flow over a flat plate requires a suitable geometry to avoid the transition to turbulence for low Reynolds numbers. Roach and Brierley [19] studied the flow over a flat plate with cylindrical leading edge of 2 mm diameter, tapered to 5° from the upper surface, to avoid instabilities and separation. Palikaras et al. [20] experimentally and numerically studied the effect of the semicircular leading edge on the transition laminar-turbulent from a flow on a flat plate without pressure gradient. They showed that this transition occurs in the presence of a pressure gradient in the region of the leading edge, resulting in the formation of the laminar bulb of separation. Several configurations were designed to avoid this phenomenon [21–26]. To avoid the influence of this disturbance, Walsh et al. [23] have designed and realized preliminary measurements from a new flat plate facility for aerodynamic research. This flat plate was consisted of a leading edge radius of 1 mm with a 5° chamfer in the intrados and adjustable positive or negative trailing edge flap deflections. The plate was made wof aluminum with 10 mm thick, 1 m long, and 0.29 m wide. Patten et al. [26] studied the design effectiveness of a new flat plate with trailing edge flap. They showed that the stagnation point anchored on the upper surface and the measurements along the flat plate were compared favorably to the Blasius profile.

This paper presents the steps to develop new laws of boundary layer profiles on a horizontal flat plate with uniform suction even in the impermeable case. In the first section, we analytically resolved the governing equations by using the integral method of Bianchini and by inserting particular solutions as Error and Lambert functions. Next, we numerically studied, by using CFD Fluent, the effects of the geometrical parameters of the flat plate (leading edge, trailing edge flap angle) on the boundary layer flow. Finally, the analytical solutions of boundary layer equations were validated with the present numerical results and the literature results in all cases with and without suctions.

2. Theoretical study

In this section, we have presented a new analytical approximation of boundary layer profiles of flow on the upper side of flat plate without and with uniform suction. Based on the approximate integral method of Bianchini and using Lambert and/or Error functions, we have achieved this solution.

2.1 Mathematical formulation

We consider a horizontal flat plate placed in incompressible, two-dimensional, steady, and laminar flow with free-stream velocity U. The x-coordinate and y-coordinate are measured from the leading edge and normal to the flat plate, respectively. In the case of permeable flat, the suction velocity v_p is oriented to y-negative (**Figure 1**).

Prandtl equations in the boundary layer are:

$$\frac{\partial u}{\partial x} + \frac{\partial v}{\partial y} = 0, \tag{1}$$

$$u\frac{\partial u}{\partial x} + v\frac{\partial u}{\partial y} = \upsilon\frac{\partial^2 u}{\partial y^2} \tag{2}$$

The boundary conditions are:

$$u(0) = 0; v(0) = v_p \tag{3}$$

$$u(\infty) = U_\infty; v(\infty) = 0 \tag{4}$$

The integration of Eqs. (1) and (2) from 0 to ∞ with the conditions Eqs. (3) and (4) gives the following integral equation:

$$U_\infty^2 \frac{d}{dx}\int_0^\infty \frac{u}{U_\infty}\left(1 - \frac{u}{U_\infty}\right)dy - U_\infty v_p = \upsilon\left(\frac{\partial u}{\partial y}\right)_{y=0} \tag{5}$$

The basic assumption of the integral method of Bianchini is to pose for the profile speed the following form:

$$\frac{u}{U_\infty} = erf\left(\frac{y}{h(x)}\right) \tag{6}$$

Figure 1.
The problem schematic.

where $erf\left(\frac{y}{h(x)}\right)$ is the Error function and $h(x)$ is the unknown scale function to be determined. The choice of the Error function has the advantage of a good approximation of the exact solution of Blasius which will be proven below.

The insertion of Eq. (6) in Eq. (5), with the conditions Eqs. (3) and (4), gives a differential equation of $h(x)$, with a boundary condition as below:

$$\begin{cases} U_\infty \alpha_2 \dfrac{dh(x)}{dx} - v_p = \dfrac{v}{h(x)}\alpha_3 \\ h(0) = 0 \end{cases} \tag{7}$$

where

$$\alpha_2 = \int_0^\infty erf(z)(1 - erf(z))dz = \frac{\sqrt{2}-1}{\sqrt{\pi}}, \tag{8}$$

$$\alpha_3 = \left(\frac{derf}{dz}\right)_{z=0} = \frac{2}{\sqrt{\pi}}. \tag{9}$$

2.2 Analytical solutions

The analytical resolution of the differential Eq. (7) depends inevitably on the boundary conditions, in particular, the value of suction velocity v_p, since we consider two cases: an impermeable flat plate when $v_p = 0$ and a porous flat with uniform suction when $v_p = -v_0 \neq 0$.

2.2.1 Case without suction ($v_p = 0$)

When replacing in Eq. (7) the value of $v_p = 0$, the differential equation of $h(x)$ is written as

$$U_\infty \alpha_2 \frac{dh(x)}{dx} = \frac{v}{h(x)}\alpha_3 \tag{10}$$

The resolution of Eq. (10) with the boundary condition in Eq. (7) leads to the solution:

$$h(x) = 2\sqrt{1+\sqrt{2}}\sqrt{\frac{vx}{U_\infty}} \tag{11}$$

So, we obtain the profile of the boundary layer velocity of the flow on the impermeable flat plate:

$$\frac{u}{U_\infty} = erf(0,32\eta) \tag{12}$$

These results enable us to determine the various characteristics of the boundary layer as the boundary layer thickness, the displacement thickness, the momentum thickness, and the friction coefficient:

$$\frac{\delta}{x} = \frac{5,66\,\delta}{\sqrt{Re_x}\,x} = \frac{5,66}{\sqrt{Re_x}};\ \frac{\delta_1}{\delta} = 0,31;\ \frac{\delta_2}{\delta} = 0,128;\ \frac{1}{2}Cf = \frac{0,36}{\sqrt{Re_x}} \tag{13}$$

2.2.2 Case with uniform suction

We considered the case of a flow on a permeable flat plate with uniform suction $v_p = -v_0 \neq 0$. To simplify the resolution of the differential Eq. (7), we imposed the particular solution:

$$h(x) = A_1 + A_2 W(g(x)) \tag{14}$$

where A_1 and A_2 are constant parameters, W is the Lambert function, and $g(x)$ is a function of x to determine.

By inserting Eq. (14) in the differential Eq. (7), we supplied the parameters of the scaling function:

$$A_1 = A_2 = \frac{b_1}{\frac{v_p}{U_\infty}}; g(x) = -\frac{1}{2v} \exp\left(-\frac{b_2}{\left(\frac{U_\infty}{v_p}\right)^2} x + b_3\right) \tag{15}$$

where

$$b_1 = -\frac{\alpha_3 v}{U_\infty} = -\frac{2v}{\sqrt{\pi} U_\infty}, b_2 = \frac{(1+\sqrt{2})\pi U_\infty}{2v}, \text{ and } b_3 = \ln(2v) - 1 \tag{16}$$

Thus, the profile of the boundary layer velocity of the flow on the permeable flat plate with uniform suction is

$$\frac{u}{U_\infty} = erf\left(y / \frac{b_1}{\frac{v_p}{U_\infty}}\left[1 + W\left(-\frac{1}{2v} \exp\left(-\frac{b_2}{\left(\frac{U_\infty}{v_p}\right)^2} x + b_3\right)\right)\right]\right) \tag{17}$$

These results enable us to determine the various characteristics of the boundary layer as the boundary layer thickness, the displacement thickness, the momentum thickness, and the parietal friction coefficient:

$$\delta(x) = 1,82 h(x); \delta_1(x) = 0,564 h(x); \delta_2(x) = 0,231 h(x); Cf(x) \mathrm{Re}_x = \frac{4}{\sqrt{\pi}} \frac{x}{h(x)} \tag{18}$$

We can rewrite this friction coefficient in the universal form of law recommended by Iglisch (1949) [11].

$$\frac{Cf(t)}{2\frac{v_p}{U_\infty}} = f\left(t = -\frac{v_p}{U_\infty}\sqrt{\mathrm{Re}_x}\right) \tag{19}$$

Thus

$$\frac{Cf(t)}{2\frac{v_p}{U_\infty}} = -\frac{1}{1 + W\left(-\frac{1}{2v} \exp\left(-\frac{(1+\sqrt{2})\pi}{2} t^2 + b_3\right)\right)} \tag{20}$$

3. Validations of the new boundary layer theories

In order to validate the new laws of boundary layer without and with suction, we studied the 2D, laminar, and incompressible flow around the flat plate by means of CFD using the software package Fluent.

3.1 CFD study of the boundary layer

Theoretically, the boundary layer equations were studied with hypothesis of zero pressure gradient flow. In order to validate and compare the numerical results with the new boundary layer laws, it is essential to avoid separation and instability of the flow. A critical part of the experiment that must be addressed is the leading edge of the flat plate. The boundary layer development is considerably influenced by the stagnation point location. A flat plate equipped with adaptable trailing edge flap ensures that the boundary layer is developed smoothly and a negligible stream-wise pressure gradient is achievable. So, the laminar separation bulb, which is one of the phenomena at the origin of the transition to turbulence, is avoided [23, 24, 26].

For this reason, we used a flat plate with a semicircular leading edge and provided with trailing edge flap deflected to an angle β = 40°. Its length and thickness are L = 0.9 m and e = 0.01 m, respectively, the flap chord l = 0.1 m. The leading edge is chamfered at an angle of 5° relative to the lower surface, and its diameter is d = 0.002 m (see **Figure 2**). This is similar to the flat plate designed by Patten et al. [26] for boundary layer research.

A judicious choice of the suction system is required to control the laminar flow [1, 27–29]. The suction area is placed in 0.1 m from the leading edge and extends to 0.7 m. It consists of holes of the same diameter d_{suc} = 0.2 mm and equidistant from 1 mm (**Figure 2**). This corresponds to a longitudinal dimensionless spacing d_p/d_{suc} = 5 (where d_p is the spacing between two successive holes), respecting the mechanical strength of the plate [30].

The quality of mesh has a great importance on the results of a numerical resolution. For this, we choose a structured and very tight mesh in the near-wall and in the area of the leading edge (Y+ = 1). This computational domain is made up of 53.970 cells in the case of the impermeable plate and 100.000 cells in the case of the

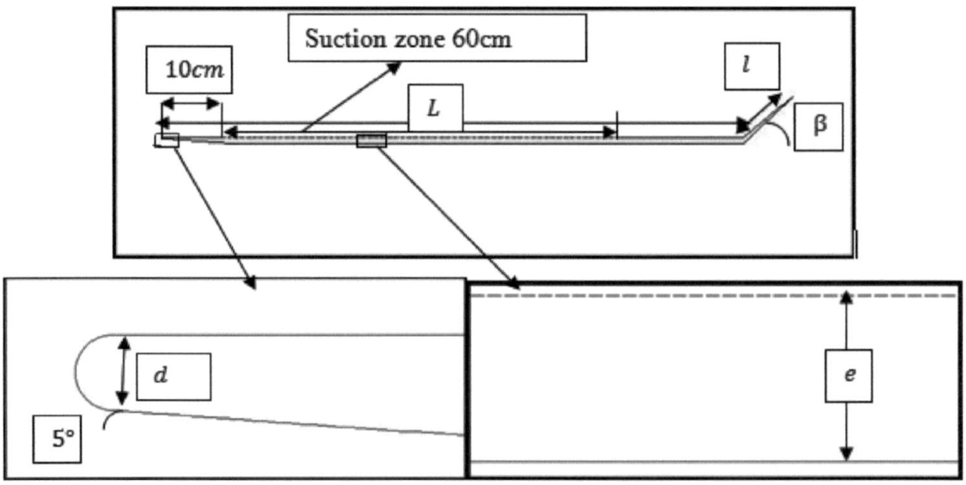

Figure 2.
Design characteristics of the flat plate.

permeable plate, because of the mesh refinement above the suction zone to account for the velocity gradient. The Navier-Stokes equations for 2D, laminar, and incompressible flow were resolved by using the finite volume method (FVM). We used the algorithm "SIMPLE" for the pressure-velocity coupling.

In the case without suction, we studied the flow with free-steam velocity $U = 5$ m/s around the flat plate to compare their boundary layers with Blasius profiles. **Figure 3** shows the effect of the trailing edge flap angle on the position of the stagnation point. For $\beta = 40°$, the stagnation point is displaced to the upper surface of plate resulting in the reduction of the separation flow at the leading edge compared to the case of $\beta = 0°$. This result is compared to those obtained in the literature [26].

In **Figure 4**, we compared the Blasius profiles with the results from the CFD of the flow boundary layer on the upper side of the impermeable flat plate for $\beta = 40°$. It is shown that the boundary layer of the flat plate at different positions favorably follows the Blasius profile. Thus, the shapes of the leading edge and the deflected trailing edge have an effect to neglect the pressure gradient in the flow of the upper side of the plate which greatly influences the formation of the boundary layer. In the continuation of this work, we select the case of the trailing edge deflected to 40°.

3.2 Validation and discussion

Many solutions were found based on the Prandtl equations such as Blasius and Schlichting [9] profiles. The differential equations of Blasius have no solutions for

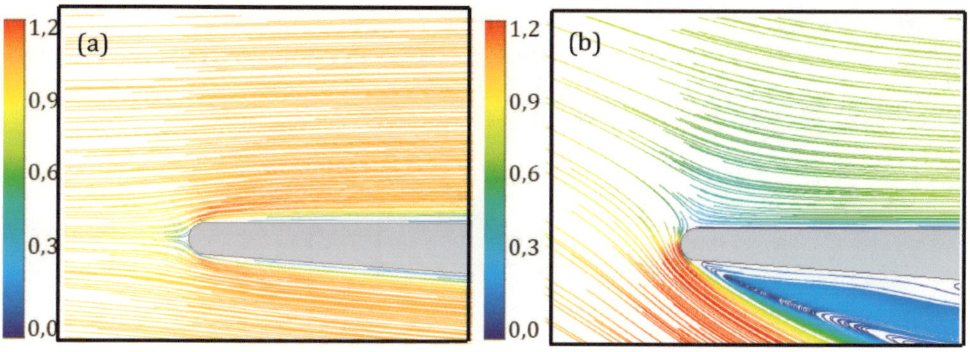

Figure 3.
Streamline velocity colored by dimensionless velocity magnitude ($(u^2 + v^2)^{0,5}/U$): (a) trailing edge flap angle $\beta = 0°$; (b) trailing edge flap angle $\beta = 40°$.

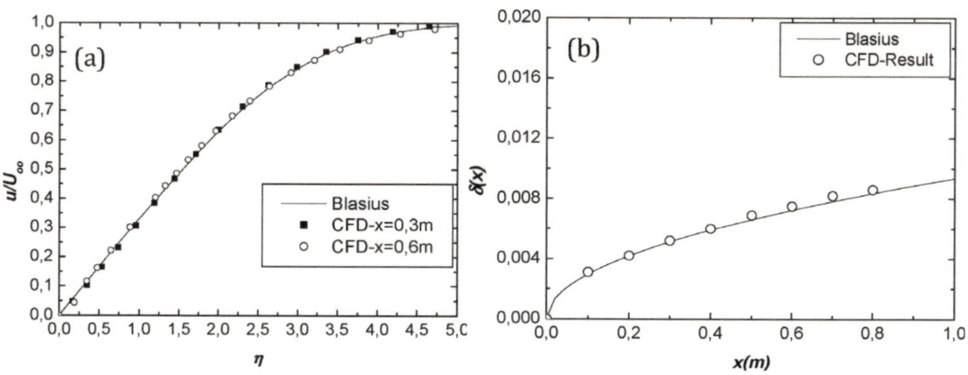

Figure 4.
Comparison between Blasius and CFD profiles for impermeable flat plate for $\beta = 40°$.

the case of uniform suction. Schlichting [9] neglected the dependency between the boundary layer velocity and the x-coordinate; this hypothesis is not acceptable in the region near the leading edge. In the case with uniform suction, Palekar and Sarma [2] found two solutions for boundary layer profiles: the first one nears to the leading edge and the second one far the leading edge. The advantage of the new solutions is that the Error function defines well the boundary layer near and far the leading edge in the cases with and without uniform suction, as well as, in the case of nonuniform suction.

Figure 5 compares the boundary layer profile defined in Eq. (12), with the CFD, and the Blasius profiles for the impermeable flat plate. This result confirms the choice of the profile shape ($erf(y/h(x))$. This shows that the Error function has the advantage of a good approximation of the boundary layer theory. We compared in **Table 1** the new characteristic parameters for the boundary layer with those from the literature. The values found by this approximation are quite comparable to Blasius and generally better than the other approximations.

Concerning the case of uniform suction ($v_p = -0.0118$ m/s), we compare the profile defined by Eq. (17) with the Palekar profiles for small and gross x. Our analytical solution verifies well both cases at once (**Figure 6**). As shown in **Figure 7**, the relation Eq. (20) well verified the universal law of friction found by Iglisch (1949) [11]. The comparison between the analytical profiles Eq. (17) with the numerical results (CFD) shows a little difference. This is caused by the nature of the parietal suction for each case, where in the theoretical study, the boundary condition at the wall is defined as a continuous suction along the plate; however, in the CFD study, the suction is discrete.

After validation of the new analytical laws, we presented the effect of the suction rate on the characteristics of the boundary layer. **Figure 8a, b**, and **c**, shows

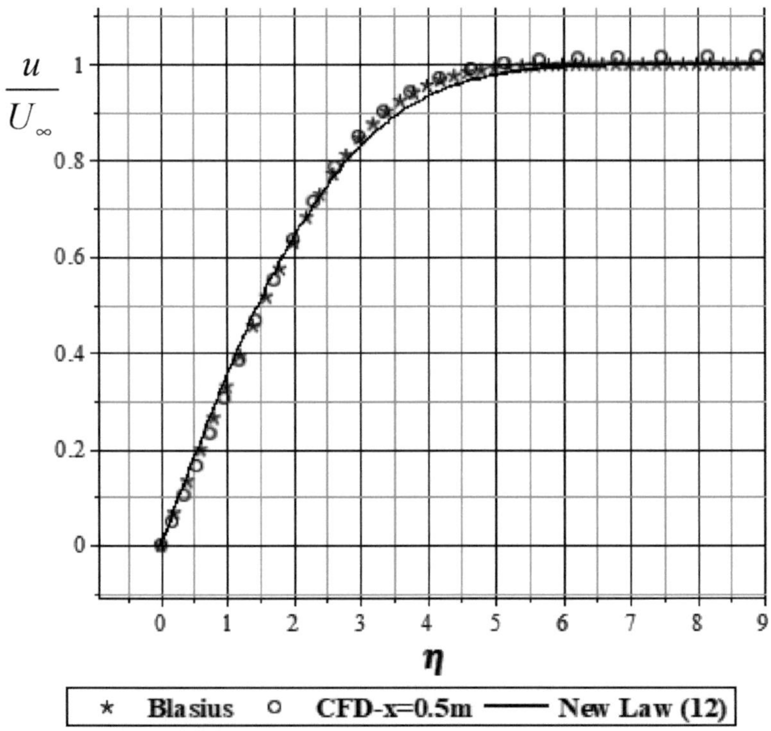

Figure 5.
Comparison of the analytic velocity profile Eq. (12) with the profile of Blasius and the profile obtained by using CFD Fluent for the impermeable flat plate.

$\frac{u}{U_\infty}$	$\delta\sqrt{\frac{Re_x}{x}}$	$\frac{\delta_1}{\delta}$	$\frac{\delta_2}{\delta}$	$\frac{Cf}{2}\sqrt{Re_x}$
$2\eta-\eta^2$	5, 4	0, 33	0, 4	0, 36
$\frac{3}{2}\eta-\frac{1}{2}\eta^3$	4, 6	0, 375	0, 139	0, 33
$2\eta-2\eta^3+\eta^4$	5, 8	0, 3	0, 121	0, 34
Blasius solution	5	0, 344	0, 132	0, 332
$erf(0,32\eta)$	5, 66	0, 31	0, 128	0, 36

Table 1.
Comparative table of characteristic parameters of boundary layer.

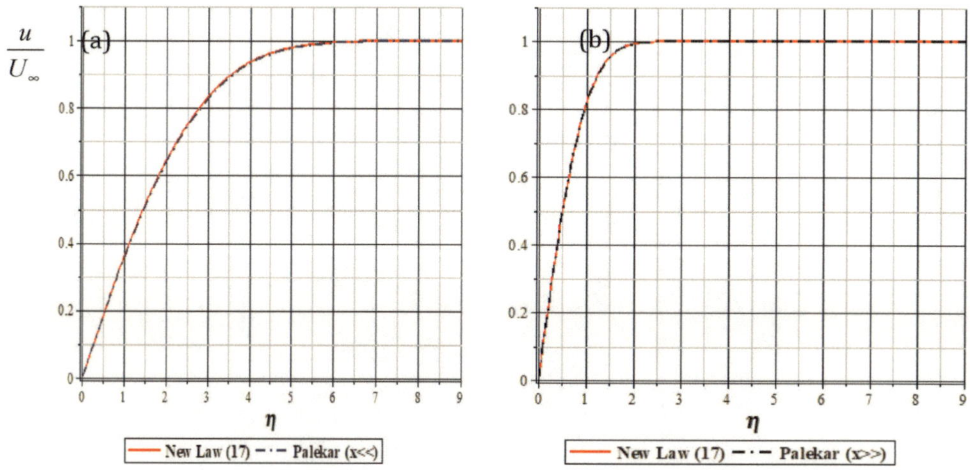

Figure 6.
Validation of the new law in the case of uniform suction with the solutions found by Palekar (1984) [2], for small x ($x = 0.001\ m\ \times = 10^{-5}$) and gross x ($x = 0.5\ m\ \times = 10^{-5}$).

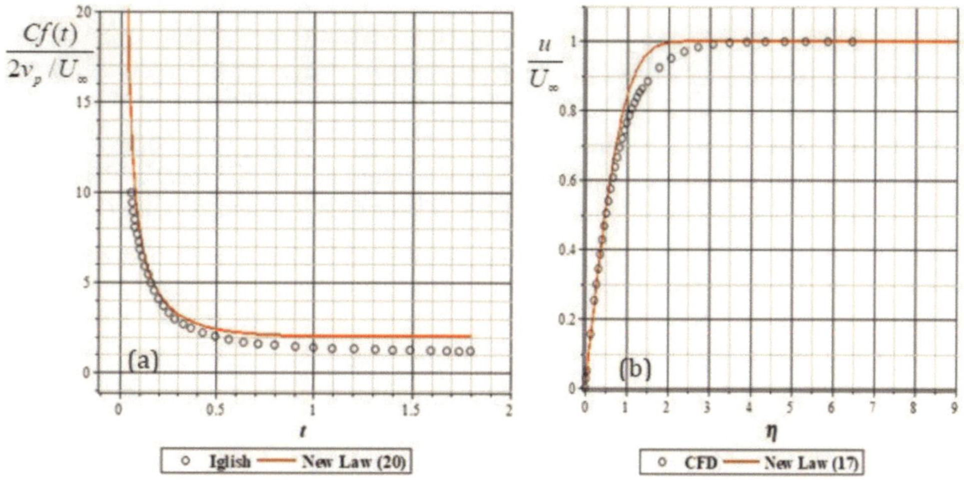

Figure 7.
Representing curves of (a) the universal law of friction $\frac{Cf}{2\frac{v_p}{U_\infty}}$ depending on t; (b) boundary layer profile (x = 0.5).

the velocity profiles, the parietal friction coefficients, and the boundary layer thicknesses for different suction rates. It is clear that when the suction rate increases, the thickness of the boundary layer decreases. As a result, the boundary

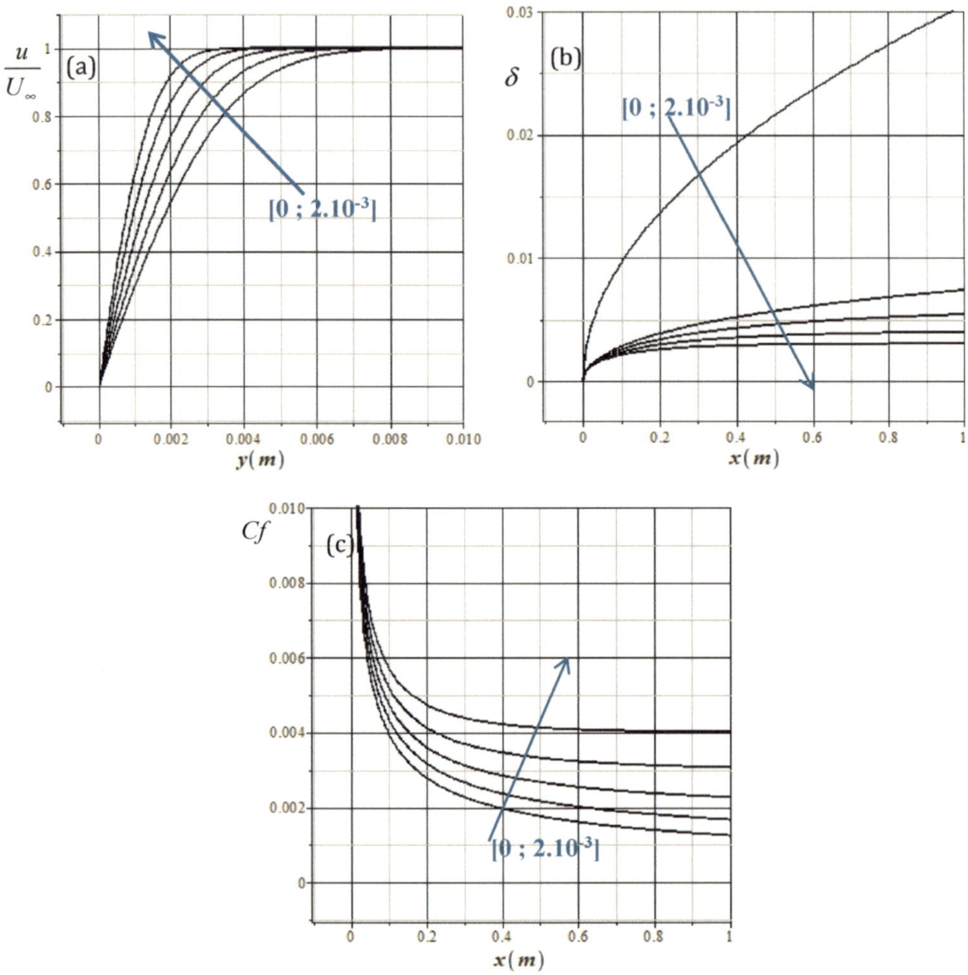

Figure 8.
Parameters of boundary layer for different values of suction rate (v_p/U = 0; 5.10^{-4}; 10^{-3}; 1,5.10^{-3}; 2.10^{-3}) for U = 5 m/s; (a) velocity profiles; (b) boundary layer thicknesses; (c) parietal friction coefficients.

layer profile flattens and the skin friction coefficient increases. This increase has no great effect on the total drag because it depends essentially on the form of drag. The contribution of the friction drag is negligible. This result is in accord with the literature.

4. Conclusion

This article has an objective to provide the analytical solutions of profile of boundary layer without and with uniform suction and to contribute to a better description of the structure of the flat plate to control the boundary layer. So, we presented the analytical resolution of the boundary layer equations by using the Bianchini integral method. This leads to new theoretical approximations, with the Error and/or Lambert functions. This result allows us to bring out the analytical expressions of several boundary layer features.

The new boundary layer theories were validated with literature results, as well as, with results obtained from numerical simulations using CFD Fluent.

Nomenclatures

Cf	skin friction coefficient
dd	diameter of the leading edge of the plate (m)
d_{asp}d$_{asp}$	thickness of the suction holes (m)
$d_p d_p$	distance between two successive holes (m)
e	plate thickness (m)
L	length of the plate (m)
l	length of the trailing edge flap of the plate (m)
Re_x	Reynolds number based on x
uu	velocity component along x (m/s)
U_∞	velocity inlet (m/s)
vv	velocity component along y (m/s)
v_p	suction velocity (m/s)
v_0	absolute value of the suction velocity (m/s)
x	longitudinal coordinate (m)
y	transversal coordinates (m)
β	trailing edge flap angle (°)
ν	kinematic viscosity of the fluid (air) (m^2/s)
δ	boundary layer thickness (m)
δ_1	displacement thickness (m)
δ_2	momentum thickness (m)
τp	parietal friction force (N/m^2)
$\eta = y\sqrt{\frac{U_\infty}{\upsilon x}}$	Blasius parameter

Author details

Chedhli Hafien[1*], Adnen Bourehla[2] and Mounir Bouzaiane[1]

1 Laboratory of Mechanics of Fluids, Faculty of Science of Tunis, Tunis Cedex, Tunisia

2 Aviation School of Borj El-Amri, Tunisia

*Address all correspondence to: chedhli.hafien@gmail.com

References

[1] Hwang D. Review of research into the concept of the microblowing technique for turbulent skin friction reduction. Progress in Aerospace Sciences. 2004;**40**:559-575

[2] Palekar MG, Sarma DP. Approximate solutions of the boundary layer equations with suction and blowing. Indian Journal of Pure and Applied Mathematics. 1984;**15**(6):671-680

[3] Yoda M, Westerweel J. Particle image velocimetry studies of a boundary layer perturbed by localized suction. Experiments in Fluids. 2001;**30**:239-245

[4] Aydin O, Kaya A. Laminar boundary layer flow over a horizontal permeable flat plate. Applied Mathematics and Computation. 2005;**161**:229-240

[5] Fang T, Liang W, Lee CF. A new solution branch for the Blasius equation —A shrinking sheet problem. Computers & Mathematics with Applications. 2008;**56**:3088-3095

[6] Thiagarajan M, Senthilkumar K. DTM-pade approximants for MHD flow with suction/blowing. Journal of Applied Fluid Mechanics. 2013;**6**(4): 537-543

[7] Asaithambi A. Numerical solution of the Blasius equation with Crocco-Wang transformation. Journal of Applied Fluid Mechanics. 2016;**9**(5):2595-2603

[8] Zheng J, Han X, Wang ZT, Li C, Zhang J. A globally convergent and closed analytical solution of the Blasius equation with beneficial applications. AIP Advances. 2017;**7**:065311

[9] Schlichting H, Bussmann K. Exakte Losungen fiir die Laminare Reibungsschicht mit Absaugung und Ausblasen, Schr. Dr. Akad. LuftForsch; **7B**(2); 1943

[10] Preston JH. The boundary layer flow over a permeable surface through which suction is applied. A.R.C. R & M 2244; 1948

[11] Iglisch R. Exact calculation of the laminar boundary layer in longitudinal flow over a flat plate with homogeneous suction. N.A.C.A. Tech. Memo 1205; 1949

[12] Bianchini A, de Sosio L, Pozzi A. Approximate solutions of the unsteady boundary layer equations. Journal of Applied Mechanics. 1976;**43**(3):396-398

[13] Kay JM. Boundary layer flow along a flat plate with uniform suction. A. R. C., R &M 2628, London: Aeronautical Research Council; 1948

[14] Savas O. An approximate compact analytical expression for the Blasius velocity profile. Communications in Nonlinear Science and Numerical Simulation. 2012;**17**:3772-3775

[15] Prand K, Dehghan M, Baharifard F. Solving a laminar boundary equation with the rational Gegenbauer functions. Applied Mathematical Modelling. 2013; **37**:851-863

[16] Marinca V, Herisanu N. The optimal homotopy asymptotic method for solving Blasius equation. Applied Mathematics and Computation. 2014; **231**:134-139

[17] Wedin H, Cherubini S, Bottaro A. Effect of plate permeability on nonlinear stability of the asymptotic suction boundary layer. Physical Review E. 2015;**92**:013022

[18] Cherubini S, Palma PD, Robinet JC. Nonlinear optimals in the asymptotic suction boundary layer: Transition thresholds and symmetry break. Physics of Fluids. 2015;**27**:034108

[19] Roach PE, Brierley DH. The influence of a turbulent free stream on zero pressure gradient transitional boundary layer development. Part 1 test cases t3a and t3b. In: ERCOFTAC Workshop, Lausanne, France; 1990

[20] Palikaras A, Yakinthos K, Goulas A. Transition on a flat plate with a semi-circular leading edge under uniform and positive shear free-stream flow. International Journal of Heat and Fluid Flow. 2002;**23**:455-470

[21] Jonas OM, Uruba V. Experiments on by-pass boundary layer transition with several turbulence length scales. In: IMechE; C557/107/99; 1999

[22] Fransson JHM, Matsubara M, Alfredsson P. Transition induced by free-stream turbulence. Journal of Fluid Mechanics. 2005;**527**:1-25

[23] Walsh EJ, Hernon D, Davies MRD, McEligot DM. Preliminary measurements from a new flat plate facility for aerodynamic research. In: 6 European Conference On Turbo Machinery; INEEL/CON-04-02301 PREPRINT; 2005

[24] Hernon D, Walsh EJ, McEligot DM. Experimental investigation into the routes to bypass transition and shear-sheltering phenomenon. Journal of Fluid Mechanics. 2007;**591**:461-479

[25] Hernon D, Walsh EJ, McEligot DM. Instantaneous fluctuation velocity and skewness distributions upstream of transition onset. International Journal of Heat and Fluid Flow. 2007;**28**:1272-1279

[26] Patten N, Young TM, Griffin P. Design and characteristics of new test facility for flat plate boundary layer research. World Academy of Science, Engineering and Technology. International Journal of Aerospace and Mechanical Engineering. 2009;**3**(10): 1183-1189

[27] Shojaefard MH, Noorpoor AR, Avanesians A, Ghaffarpour M. Numerical investigation of flow control by suction and injection on a subsonic airfoil. American Journal of Applied Sciences. 2005;**2**(10):1474-1480

[28] Iborra P, Rabeau F, Saalbach C. Contrôle de décollement sur rampe par fente synthétique pulsée. BEI EP 2007/2008 ENSEEIHT-Département Hydraulique; 2008

[29] Genç MS, Kaynak Ü, Yapici H. Performance of transition model for predicting low Re aerofoil flows without/with single and simultaneous blowing and suction. European Journal of Mechanics—B/Fluids. 2011;**30**: 218-235

[30] Mendez MS. Simulation Numerique Et Modelisation De L'ecoulement Autour Des Parois Multi-Perforees [Thesis]. University Montpellier Ii; Cerfacs: Th/Cfd/07/104; 2007

Section 4

Aerospace Applications

Chapter 7

High Entropy Alloys for Aerospace Applications

Modupeola Dada, Patricia Popoola, Samson Adeosun and Ntombi Mathe

Abstract

In the aerospace industry, materials used as modern engine components must be able to withstand extreme operating temperatures, creep, fatigue crack growth and translational movements of parts at high speed. Therefore, the parts produced must be lightweight and have good elevated-temperature strength, fatigue, resistant to chemical degradation, wear and oxidation resistance. High entropy alloys (HEAs) characterize the cutting edge of high-performance materials. These alloys are materials with complex compositions of multiple elements and striking characteristics in contrast to conventional alloys; their high configuration entropy mixing is more stable at elevated temperatures. This attribute allows suitable alloying elements to increase the properties of the materials based on four core effects , which gives tremendous possibilities as potential structural materials in jet engine applications. Researchers fabricate most of these materials using formative manufacturing technologies; arc melting. However, the challenges of heating the elements together have the tendency to form hypoeutectic that separates itself from the rest of the elements and defects reported are introduced during the casting process. Nevertheless, Laser Engineering Net Shaping (LENS™) and Selective Laser Melting (SLM); a powder-based laser additive manufacturing process offers versatility, accuracy in geometry and fabrication of three-dimensional dense structures layer by layer avoiding production errors.

Keywords: high entropy alloys, laser additive manufacturing, LENS, SLM, laser scan strategies

1. Introduction

There are three major components of the turbine engine; the compressor, the combustor, the turbine blade and the nozzle. In recent years, the aeronautic trade demands the advancements of new material for the aero-engine components characterized by thrust, weight, safety, fuel utilization, life cycle costs and environmental necessities [1]. Contemporary innovative advances and evolution in the aerospace industry require improvement and application of structural materials that would provide higher performance and will be cost-effective in fabrication and maintenance compared with existing parts. The choice of material relies upon the working conditions and an ideal alloy that could withstand an environment with extreme temperatures while lightweight. Consequently, the aero engine material distribution comprises steels, titanium alloys, nickel superalloys, aluminum alloys and more recently high entropy alloys [2].

The avionic trade utilizes low-alloy steels, maraging steels and highly alloyed secondary stainless steels in commercial and military aircraft for their resistance from erosion, oxidation and the capacity to withstand high temperatures. However, a lot of strength is required for the steel to reduce the weight of its parts and the fracture mechanism of the material are not well understood reducing its usage for aerospace applications. Aluminum alloys were at a point fascinating to use on aero engines due to its low density but the inability to withstand elevated temperatures limited its applications. On the other hand, Ti-6Al-4V and other titanium alloys including nickel superalloys, chromium super alloys, tantalum and rare earth metallic alloys are used for their ability to withstand high temperatures, low density and lightweight [3]. However, failure of these materials has been inevitable because of extreme operating temperatures, cyclic and translational movement of the part, mode of fabrication of parts and nature of the material. Material improvements and technological advancements in the mode of synthesis led to the discovery of high entropy alloys.

Some key benefits of high entropy alloys are:

- With a hoisted strength-to-weight ratio, good oxidation resistance, fatigue resistance, hot consumption opposition, elevated temperature strength, lightweight, wear and creep resistance. High entropy alloys are excellent materials for compressors, combustion chambers, exhaust nozzle and gas turbine case applications within the gas turbine engine [4].

- With excellent creep resistance, less radioactive waste, elevated temperature strength and ion irradiation resistance, high entropy alloys are great auxiliary materials used for nuclear reactors applications [5].

- With excellent mechanical, electrical, electrochemical and anticorrosion properties, high entropy alloys are alternative alloys for biomedical applications [6].

2. High entropy alloys as aero-engine materials

High entropy alloys (HEAs) are alloys with at least five metallic components and every one of these components has a molar atomic concentration somewhere between 5 and 35% [4]. Reports on most HEAs show the amalgams contain a basic face-centered cubic (FCC) or body-centered cubic (BCC) or hexagonal closed pack (HCP) solid solution phases without intermetallic phases because of their high-entropy impact [5]. These solid solution phases empower high entropy alloys combinations to have remarkable properties, for example, increased hardness, high fracture strength, yield stress, and plastic strain. HEAs exhibit good ductility, they have a superb work hardenability and high-temperature oxidation resistance [6]. They contain particular attractive magnetic properties, have high wear resistance and exhibit good erosion opposition [7]. As a result of these properties, HEAs are remarkable refractory materials, fatigue resistant materials, and have a corrosion-resistant surface layers and diffusion obstruction layers for various structural jet engine applications. However, most HEAs were fabricated using conventional techniques and an enhancement in the manufacturing process of the alloys will upgrade the mechanical properties of these alloys.

Other materials used for aerospace applications are:

Material	Properties	Short comings
Metals: ferrous metals, e.g., stainless 15-5PH and non-ferrous metals, e.g., aluminum, e.g., aluminum 6062, aluminum 7050, aluminum 7075, Al-Li and magnesium	Good electric and thermal conductivity, high strength and corrosion resistance, light weight and ductility, hardness, high heat resistance	Magnesium is exceedingly combustible; steel is non-receptive to warm treatment. There is a surprising expense of delivering aluminum parts and thin aluminum cannot be compound bent while conveying load
Heat resistant super alloys, e.g., nickel 718, Ti6Al4V, Ti-5553, TiAl-based alloys	Elevated temperature strength and corrosion resistance	Titanium and nickel alloys and some ceramic materials are grouped under heat-resistant alloys; in any case, they are hard to machine meaning shorter life expectancy and decreased process security
Composite materials	Low weight, high tensile strength and compression resistance	The most common composite material used in the aerospace is fiberglass and it is expensive to fix while resins used as matrix discharges poisonous fumes

When designing high entropy alloys, the components in the blend need not be equivalent or near equal. The structure does not separate a minor or major element in its composition rather, the major consideration for designing high entropy alloys amalgams is; the elements in the mix ought to be at least 5 in number and additionally, their atomic concentrations should be between 5 and 35%. Higher configurational entropy is achieved by having no less than five elements in the composition because an increment in the number of components increases the mixing entropy and best explains the high entropy effect which is an important factor to the designing of high entropy alloys [8–10].

$$N_{majorelements} \geq 5 \text{ at.\%} \leq 35 \text{ at.\%} \tag{1}$$

$$N_{minorelements} \geq 0 \geq 5 \text{ at.\%} \tag{2}$$

When the Gibbs free energy (ΔG_{mix}) is at its base, the system is said to be at equilibrium.

$$\Delta G_{mix} = \Delta H_{mix} - T\Delta S_{mix} \tag{3}$$

where the enthalpy of mixing is the ΔH_{mix}, ΔS_{mix} is the entropy of mixing and T is the temperature. From the Boltzmann hypothesis on the entropy of mixing, the molar configuration entropy design (ΔS_{Conf}) gives more prominent outcomes of forming a multi-element solid solution phase through statistical thermodynamics determined by:

$$\Delta S_{conf} = -KInw = -RIn\frac{1}{n} = -RInn\,\Delta S_{mix} = R\,In_{(n)} \tag{4}$$

$$\Delta S_{conf} = R\,In_{(n)} \tag{5}$$

where ΔS_{conf} is the molar concentration, K is the Boltzmann constant, w is the thermodynamic probability, R is the gas constant (8.134 J/kmol) and n is the number of elements in the composition [8].

High entropy alloys form stable solid solutions because of the crystal structure of the elements which is unaltered when different elements are included. Furthermore, they are stable when the chemical components remain in a single homogeneous phase. This happens when the elements in the composition are firmly packed together on the periodic table.

There is a relationship between the entropy of fusion and the phase transition metals used in the compositional design of high entropy alloys. At the point when the number of elements in a composition is increased, the framework will be progressively stable.

For instance, from Eq. (5),

$$\Delta S_{conf} = R\,In_{(n)}$$

When the number of elements in a system $n = 5$, $\Delta S_{conf} = 1.61R$ where R is melting point value, when the number increases, the value of R increases. Thus, increasing the number of elements in a composition increases the mixing entropy which increases the stability of the alloy system.

Cantor et al. [11] demonstrated the number of alloys that have been studied including unitary, binary, ternary and high entropy alloy in a system with the total number of different possible alloys N as;

$$N = \left(\frac{100}{x}\right)^{c-1} \tag{6}$$

With 60 elements in the alloying range of the periodic table, at a material specification of 1, he gave the conservative number of possible alloys design as $\approx 10^{177}$.

2.1 Properties and production techniques of high entropy alloys

The property, microstructure and design of high entropy alloys are dependent on some core effects, the phase composition and technique of fabrication respectively [12].

2.1.1 Core effects

2.1.1.1 High configurational entropy

The high configurational entropy impact hinders the phase transformation influencing the thermodynamics of the system yet builds the formation of solid solutions. Despite having numerous meta-stable states, the equilibrium state has the most reduced free energy of mixing in solids from the second law of thermodynamics. This suggests that combining five or more components in equimolar or near equimolar concentrations may bolster two phases; the solid solution phase and the intermetallic phase. Intermetallic phase is a stoichiometric compound with super-lattices that give the high entropy alloy amalgams ominous properties while single-phase solid solution phases show a system with a total blend of elements forming simple (BCC, FCC, HCP) crystal structures over intermetallic compounds. The presence of a prolonged range order isolates the intermetallic phase for the solid solution phases. At elevated temperatures the higher the number of elements in a disordered state, the less the possibility of forming intermetallic phases.

This uses the Boltzmann equation:

$$(S = k\text{In}(N)) \tag{7}$$

To show the configurational entropy S, of an ideal solution with N, number of elements as it only regards configurational entropy each at an equimolar concentration and k is the Boltzmann constant however, this hypothesis presumes that the HEA studied constitute random sampling of all HEA systems which is not hypothetically valid.

2.1.1.2 Sluggish diffusion

Sluggish diffusion involves the kinetics of the system; low diffusion rate, increases thermal stability, an increase in recrystallization temperature, slows down grain growth, slows down phase separation and improves creep resistance which might benefit the microstructure. The presence of nanocrystals in as-cast material and amorphous materials in sputter-deposited thin films and high recrystallization temperatures enlivened the sluggish diffusion theory. Cross-diffusion happens when the focus angle of one element prompts or changes the motion of another element. This occurs when one component changes the chemical attributes of other elements in the alloy system. Increasing the number of the composition of the elements in an alloy may make the diffusion become sluggish and reduce the temperature of the system.

2.1.1.3 Lattice distortion

Lattice distortion manages the structure of the high entropy alloy system. A noteworthy contrast in the atomic radii; the movement of large and small atoms causes' lattice strain making the entire framework to have a distorted lattice. Lattice distortion prompts solid solution strengthening by restricting dislocation, the displacement occurring at each lattice spot relies on the atom dwelling at that spot.

2.1.1.4 Cocktail effect

Cocktail effect affects the properties of the system designed. It includes the compositional mixtures of elements where the consequence of the blend is both unpredictable and unexpected because of the distinctive properties the individual element provides. Adding an element to the mixture with properties realized will enhance the combination through the cocktail effect. High entropy alloys may exhibit properties dependent on the reaction between elements in the system. On the off chance that a high-temperature high entropy alloy is desired, elements with elevated temperature strength ought to be used.

2.1.2 Phase formation

In recent times, a few techniques for anticipating the phase(s) high entropy alloys will possess have emerged. The phases have been predicted most often, however not continually, using the calculation of phase diagram [13], the valence electron concentration or utilizing the thermodynamic and geometry effect.

According to Gibbs rule:

$$P = C + 2 - F \tag{8}$$

where C is the number of elements in the system, F is the degree of freedom and P is the maximum number of phases at equilibrium. This standard proposes that high entropy alloys can exhibit multiple phases, nonetheless, high entropy alloys (HEAs) are usually a single phase or double phase system but rarely having multiple phases regardless of containing multiple elements. Solid solution high entropy alloys form FCC, BCC or HCP phases due to their mixing entropies. BCC structured HEAs have high yield strengths, low ductility, limited plasticity and are brittle while the FCC structured HEAs have a low yield strength, inferior cast ability, compositional segregation high plasticity and ductility [14]. The ductility of HEAs decreases as the yield stress and comprehensive strength increases and the blend of

BCC and FCC phase produce mechanical properties with high strength and good ductility producing balanced alloys [15]. However, the combination of more BCC elements will show more BCC phases while combinations of elements with more FCC elements will show more FCC phases. Although entropy is not the only criteria for phase formation, both entropy and enthalpy must be considered. The crystal structures of elements used and the number of times the elements are used in an alloy system influences the phases found in that alloy system [16]. The hardness and yield strength of HEAs with FCC phases is smaller than the BCC phase; therefore, hardness increases due to the increase in the BCC phase. The BCC phase is more grounded than the FCC phase because of the structure and solution hardening and sometimes adding an element to the mixture can change the phase completely from BCC to FCC and vice versa [17].

2.1.3 Fabrication techniques

Preparation method of HEAs can be divided into three major routes Liquid mixing, solid mixing and gaseous mixing. The liquid mixing includes arc melting, electric resistance melting, inductive melting, Bridgman solidification and laser additive manufacturing [18]. In research, most HEAs were manufactured using Arc melting, which occurs in a vacuum sealed argon environment where the molten alloy is cast. The alloys to be fabricated are liquefied using a vacuum arc melter. The melter is fitted with a button-crucible. Melting is accomplished using a consumable tungsten electrode utilizing metal pellets as a charge striking the arc. A turbo-molecular and roughing pump is then used to pump the chamber to obtain a pressure of about 3×10^{-4} Torr [19]. Argon is filled in the chamber to reduce the pressure a little facilitating the plasma formation when the arc strikes. Then the melt pool is stirred by the plasma through the convention. Then the process is repeated several times to achieve homogeneity of the composition.

In any case, the challenges of heating the components together have the tendency to form a hypoeutectic that isolates itself from the rest of the elements due to slow cooling rates, the shape and sizes of bulk ingots are limited and fabrication of high entropy alloys also in bulk using this technique is relatively expensive. The solid mixing route involves mechanical alloying and subsequent consolidation process. Some studies have shown that mechanical alloying produces homogenous and stable nano-crystalline microstructure. While the gas mixing route includes molecular beam epitaxy, sputter deposition, pulse-laser deposition (PLD), vapor phase deposition and atomic layer deposition [20].

3. Laser additive manufacturing (LAM)

In recent times, the era of manufacturing technologies are automated, mechanized and computer integrated. Thus, additive manufacturing (AM) is a preferred alternative to conventional manufacturing technologies. AM enables industries to create products utilizing fewer parts and fabricate items that are less vulnerable to mileage pores and blowholes. It reduces new product cost by 70% and promoting time by 90% by utilizing the rapid prototyping and related assembling techniques [21]. Once the shape and dimensional resistance of a component or product are made as an automated 3-D image, a solid reproduction will be created in hours anywhere in the world. AM reduces life-cycle impacts and the heaviness of the final product. AM is versatile, flexible and customizable making it a preferred choice by most sectors of production. There is no need for storage as AM parts can be made and on demand from a computer-aided design file and along these lines, there is no compelling reason to change the production line to make one part. The AM parts are fabricated layer by layer reducing excesses [22] while human production errors are insignificant. More complex parts are produced in shorter time spans and it likewise guarantees higher product quality since parts created are without residual porosity. Thus, additive manufacturing can be solid, liquid or powder-based. The powder-based processes are better utilized in lieu of other processes because additive manufacturing of segments using a laser and powder together helps create complex structures assuring quality and strength to the finished parts. AM technologies of solid structures are realized by the successive deposition of layers of flowing powders, making the powder-base additive manufacturing methods practical and attractive.

Additive manufacturing via powder based melting is a technique utilized in most metal rapid frameworks which makes use of the continuous supply of metallic materials in powder shape and an energy source, dissolving the material while forming a melt pool which solidifies rapidly into metal layers. This rapid solidification or high cooling rate will produce fine microstructures making the final part fabricated have enhanced mechanical properties. Laser additive manufacturing via powder based melting process includes the Laser Engineering Net Shaping (LENS) [23] and selective laser melting (SLM) [24]. The SLM uses a cold powder bed technique while LENS uses a blown powder method by a laser beam through nozzles for particle deposition. SLM and LENS liquefying technique is adaptable, achieves accuracy in geometry and there is a better tendency to form fine grains, non-equilibrium phases and new chemical compounds in both SLM and LENS technique which results in improved mechanical properties of the material with minimal/zero defects.

3.1 Manufacturing processing of high entropy alloys using laser additive manufacturing

The fabrication of high entropy alloy Al-Co-Cr-Cu-Fe-Ni using laser additive manufacturing will be discussed. It should be noted that this is the most studied high entropy alloy system with limited information on the fabrication of this system via LAM technique; however, this system is in great proportion an extension of the common superalloys used as aero-engine materials.

In view of this, the powder characterization of the high entropy alloy Al-Co-Cr-Cu-Fe-Ni is achieved to check the morphology of the powders used.

3.1.1 Laser Engineering Net Shaping method (LENS™)

LENs substrates can be made out of 25 × 25 × 4.5 mm stainless steels plate were subjected to a compressor used to apply pressure jet of air to blast an abrasive material, etching the surface of the plates in a process called sandblasting, the process was done in an enclosed cabinet designed to contain and recycle the abrasive grit at a high air pressure to make the blasting faster. A small nozzle size helped the blast make a fine and uniform pattern at a close blasting distance with a slighted blast angle to create the desired effect.

Afterwards, continuous wave Nd:YAG laser processing system fitted with an off-axis nozzle with a dual hopper plasma spray powder feeder system is used to deposit the alloy system. The laser and powder stream move over the surface to create layers; two or more layers can be made to create a three-dimensional deposit. The LENS process then begins with a computer-aided design [25] file transferred to the laser Optomec system which in turn slices the information on the file into layers of the desired height. The CAD file is converted into a stereolithography file and parameters such as the hatch space and layer rotation are set. This stereo-lithography file is then converted to a motor control file and the travel speed is set.

The laser power is set and the federate is also set. Once all parameters are set, the process can begin automatically. The powders must be spherical in shape to flow smoothly through the hoppers while deposition takes place in an argon-filled chamber. The high entropy alloys fabricated is deposited on the substrate as the oxygen level of the chamber is constantly monitored.

3.2 Selective laser melting (SLM)

SLM is a powder bed fusion process [26]. It uses a laser beam which melts and then fuses the metal powders together as a thin layer of powder is deposited over the substrate plate then the laser beam fuses the powder particles selectively as dictated by the computer aided design data. Process parameters must be taken into consideration in order to fabricate a defect-free-part [27, 28]. The process parameters are laser power, laser scan speed, hatch distance, hatch overlaps, hatch style etc. and these parameters all affect the mechanical properties and influences the microstructures of the parts [29, 30].

3.3 Influence of laser additive manufacturing processing parameters on high entropy alloys

3.3.1 Process parameters

3.3.1.1 Laser power

Light amplification by stimulated emission of radiation is simply a device that generates an intense beam of coherent monochromatic light by stimulated emissions of photons from excited atoms or molecules. Lasers can be classified as; gas lasers, diode lasers, liquid (dye) lasers, fiber lasers and solid state lasers. The rate of energy input with respect to time is called the laser power. The intensity of the laser beam increases with an increase in time, therefore, to know the influence of the laser power on high entropy alloys, different laser power needs to be observed.

For instance, the influence of laser power on high entropy alloy CrMnFeCoNi deposited via laser melting deposition was studied by Xiang et al. [31]. The authors observed that the laser power influences the densification behavior of the alloy. They also observed that by changing the laser power, the proportion of equiaxed and columnar grains could be adjusted which affects the solidification and heat flux direction of the process.

3.3.1.2 Laser scan speed

Laser scan speed is the velocity of deposition carried out by the laser beam along the track created. It is the time rate at which the deposition is created when the laser beam is passed along the surface of the substrate is called the laser scan speed. Zhang et al. [32] observed that decreasing the scanning speed leads to a higher temperature of the melt pool. The laser scan speed offers the laser powers enough heating energy to melt the high entropy alloy powders and slower scan speed ensures a longer period to melt the powder layer completely. Therefore, the laser scan speed and the laser power will determine the energy density within the melt pool [33].

3.3.1.3 Laser beam diameter

The length at which the laser beam covers a focal distance in millimeters while creating a layer is called the beam size or beam diameter. The beam creates a melt pool as it moves along the track with an oval, thus, the major axis of the melt pool created is dependent on the scan speed. A decrease in the beam diameter increases the energy density which leads to a deeper depth at a constant powder feed rate.

3.3.1.4 Powder feed rate

During laser deposition, the high entropy alloy powders are carried through a feed tube by a carrier gas usually argon at a speed called the powder feed rate. The thickness of the layers is directly proportional to an increase in the powder feed rate. However, deposition of thick layers may result in a poor bond between layers as well as a high energy consumption which negatively increases the thermal stress and distortion of the high entropy alloy component.

3.3.1.5 Hatch spacing

Hatch spacing refers to the distance and overlap between two consecutive scan vectors. An overlap is required between the successive hatch lines to avoid pores and the spacing is usually less than the beam diameter. Zhou et al. [34] studied the influence of hatch spacing on high entropy alloy $Al_{0.5}CoCrFeNi$ prepared by selective laser melting (SLM) and the authors reported that the hatch spacing influences the relative density of the alloy as the relative density increases with an increase in the energy density. Notably, the porosity decreases with an increase in the hatch spacing and vice versa.

3.3.1.6 Energy density

This is also known as the powder density and it is the energy responsible for the melting of the powder on the substrate, therefore, the height of a single layer is dependent on the energy density. The energy density is directly proportional to the dilution; therefore, when the energy density is low, the dilution is low and no fusion bond can be formed amongst the high entropy alloy system.

$$E = \frac{P}{vD}\left(J/mm^2\right) \tag{9}$$

where D is the laser beam diameter, P is the laser power and v is the scanning speed, respectively.

3.4 Laser scan strategies on the morphology and formation of high entropy alloys

The laser scan strategies are used to reduce residual thermal stresses and fill a single cross section that can be subdivided into smaller sectors with scan lines. The lines can follow patterns such as spiral, zigzag, parallel, chessboard or paintbrush. When the laser scan speed is reduced thermal gradients and solidification may lead to cracks, however, when the scan speed is increased, the power has to be increased, therefore, knowing the right scan strategy to use in fabricating high entropy alloys is important to achieve a homogeneous system.

The type of laser and the process parameters of a laser additive manufacturing technique are not included as the scan strategies. The scan strategies show a pattern that influences independent variables during the LAM process, therefore, the scan strategies must first be defined before another parameter optimization is achieved.

Scan strategies can be divided into the layer and vector scan strategies and these strategies not only control the properties of the material but also are an important factor used to control the grain location and texture of the high entropy alloy microstructure [35].

3.4.1 Helix vector scan strategy

Helix vector scan strategy is most suitable for producing complex parts and it reduces deformation caused by steep thermal gradients in the parts produced. A Voronoi diagram is used to build each layer and a tool path algorithm applies to the diagram and generates the recursive helix scan path for every layer (**Figure 1**).

3.4.2 Island scan strategy

This is a strategy that tries to remove thermal residual stresses and this is achieved by putting separating exposed areas in a track into smaller sections called islands and this is usually 5 × 5 mm by default. The islands are then scanned in a random sequence with short scan tracks eliminating localized heating of the larger sections and subsequently reducing the thermal gradients and residual stresses (**Figure 2**).

3.4.3 Layer scan strategy

Layer scan strategies comprise an orthogonal scan strategy and inter-layer stagger strategy. An orthogonal scan strategy is used to reduce porosity and stresses building up along the scan track by changing the direction of the scan after each layer is built. This is achieved when consecutive layers are scanned orthogonally to each other. The inter-layer or knitting strategy is used to repair defects observed in previously scanned layers through overlapping. The defects are corrected by melting all the powder in the overlapping zone causing a strong bond between the layers (**Figure 3**).

3.4.4 Vector scan strategies

The vector scan strategy consists of the progressive and ‘raster scan strategies. The ‘raster scan strategy alternates the vector track after every scan. The laser scans

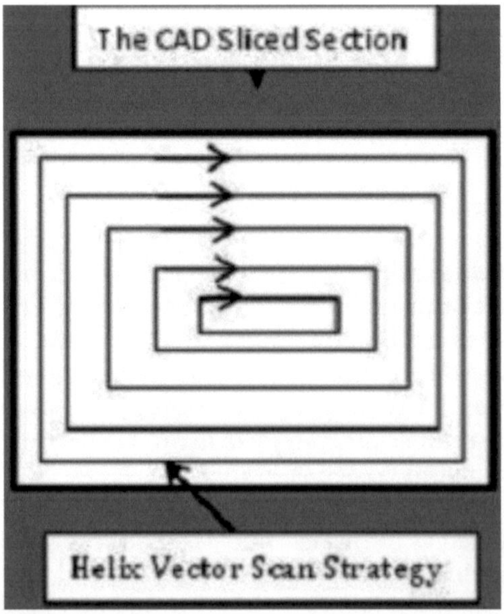

Figure 1.
Helix scan strategy; adapted from [36].

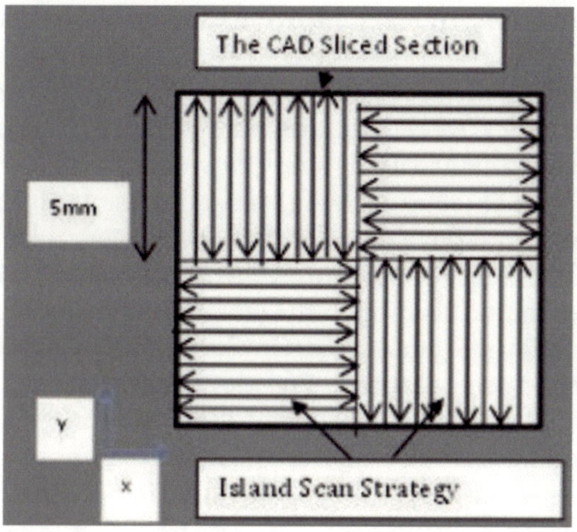

Figure 2.
Island strategy; adapted from [37].

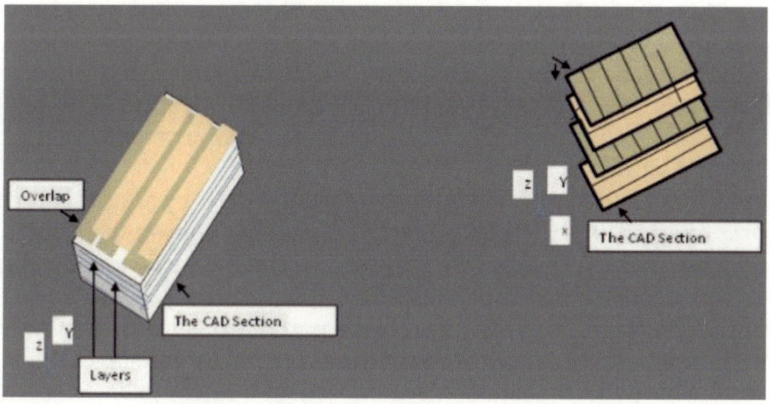

Figure 3.
Layer scan strategies; adapted from [38].

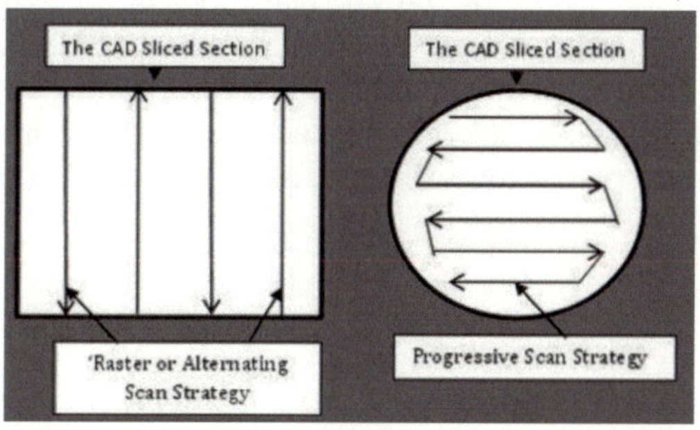

Figure 4.
Vector scan strategies; adapted from [39, 40].

from the beginning to the end of a vector before moving to the next vector beginning with the next vector at close range to the end of the previous vector. While the progressive scan strategy as the word progressive states is a scan strategy that does not stop but continues from one vector to another (**Figure 4**).

3.5 Benefits and limitations of laser additive high entropy alloys

Aero engines comprise different parts and those parts are composed of several materials; aluminum alloys, steels, titanium alloys, nickel superalloys, ceramics, composites and intermetallics to name a few, however, most of these materials have limiting properties. High entropy alloys fabricated using laser additive manufacturing through research and development show promising properties; elevated temperature strength, oxidation resistance, favorable compressive yield strength, advantages over other materials used in the jet engines despite its challenges [41].

The high entropy alloy system Al_x-Co-Cr-Cu-Fe-Ni fabricated at $x = 0.5$ exhibits high strength at elevated temperatures. The aluminum content in the high entropy system influences the crystal structure of the alloy. At reduced aluminum content, a ductile FCC phase will be formed which is resistant to changes at high temperatures and the strengthening is as a result of the solid solution phases. The wear resistance of the alloy acts independently with respect to its hardness [42]. As opposed to some alloys that the resistance to wear is directly proportioned to its hardness, this alloy system has been reported to having a high wear resistance despite its reduced hardness and this is attributed to the surface hardening of the ductile FCC phase. At low aluminum contents, delamination wear is observed while at high aluminum content, oxidation wear is observed.

Although this high entropy system shows variation in its corrosion properties from favorable to not favorable in both NaCl and H_2SO_4 solutions it has been reported to be susceptible to pitting corrosion in chloride environments, which is increased by anodizing in H_2SO_4.

The aluminum and chromium content in the high entropy alloy system has shown to improve the oxidation properties of the alloy. Aluminum achieves this by creating a protective aluminum oxide (Al_2O_3) layer on the surface while chromium also creates a protective chromium oxide (Cr_2O_3) layer on the surface [43].

The fatigue resistance of the alloy has been reported to be favorable between 540 and 950 MPa. However, there is a need to improve the fatigue resistance of the alloy as recent studies have shown that $Al_{0.5}$CoCrCuFeNi high entropy alloy is sensitive to defects, such as micro-cracks, introduced using the conventional manufacturing techniques [44]. These manufacturing defects arise and contribute to a reduced fatigue life of the material and an increase in the cost of reproduction, therefore, the removal of these defects and an increase in the fatigue resistance of the material will cause improvements of the technology of production. SLM and LENS melting technique are versatile and achieve accuracy in geometry. SLM uses a powder bed and LENS uses a blown powder method by the laser beam. Formation of fine grains, non-equilibrium phases and new chemical compounds result in improved mechanical properties

4. Conclusion

High entropy alloys (HEAs) possesses superior mechanical, thermal and oxidation properties exceeding that of pure metals. Attributed to the core effects; high mixing entropy, lattice distortion, slow diffusion and cocktail effect. HEAs have outstanding strength is comparable to some metallic glasses and that of structural ceramics attributed to the simple solid solutions they form. Valence electron concentration, CALPHAD and using the thermodynamic and geometry effect are suggested means of discovering the phase HEAs.

High entropy alloys have shown good wear and corrosion resistance with their thermal conductivity lower than that of pure metals. The alloys have remarkable superconductivity and have been reported to be applicable in high temperature and low-density refractory for the aero engine components. However, defects have been reported to limit the fatigue resistance of high entropy alloys using arc melting; the most widely used technique of fabricating high entropy alloy until recently. Many techniques have been substituted over the conventional process of manufacturing high entropy alloys; mechanical alloying, sputter deposition, molecular beam epitaxy (MBE), pulse-laser deposition (PLD), atomic layer deposition (ALD) and vapor phase deposition. However, none of these techniques is versatile, flexible and customizable. None use a computer-aided design file eliminating the need to change the production line just to make one part. None of the techniques mentioned above is built layer by layer reducing excesses while human production errors are minimal. None have more complex parts produced in shorter time frames and which assures higher product quality because parts developed are without residual porosity than the laser additive manufacturing technique.

Therefore, there are limitless possibilities in using high entropy alloys fabricated using laser additive manufacturing for aero engine applications. Not only are high entropy alloys similar to nickel-based superalloys currently in use but also a cheaper alternative.

Author details

Modupeola Dada[1*], Patricia Popoola[1], Samson Adeosun[2] and Ntombi Mathe[3]

1 Tshwane University of Technology, Pretoria, South Africa

2 University of Lagos, Akoka, Nigeria

3 Council for Scientific and Industrial Research, Gauteng, South Africa

*Address all correspondence to: dadadupeola@gmail.com

References

[1] Smith K. Aircraft propulsion and gas turbine engines—2nd Edition AF El-Sayed CRC Press, Taylor & Francis Group, 6000 Broken Sound Parkway NW, Suite 300, Boca Raton, FL, 33487-2742, USA. 2017. Distributed by Taylor & Francis Group, 2 Park Square, Milton Park, Abingdon, OX14 4RN, UK. 1447pp. Illustrated. £130. (20% discount available to RAeS members via www.crcpress.com using AKQ07 promotion code). ISBN 978-1-4665-9516-3. The Aeronautical Journal. 2018;**122**(1251):854-855

[2] Alderliesten R. Introduction to Aerospace Structures and Materials. Netherlands; 2018. pp. 41-58

[3] Rana S, Fangueiro R. Advanced Composite Materials for Aerospace Engineering: Processing, Properties and Applications. United States of America: Woodhead Publishing; 2016. pp. 1-15

[4] Yeh JW et al. Nanostructured high-entropy alloys with multiple principal elements: Novel alloy design concepts and outcomes. Advanced Engineering Materials. 2004;**6**(5):299-303

[5] Tong C-J et al. Microstructure characterization of Al_xCoCrCuFeNi high-entropy alloy system with multiprincipal elements. Metallurgical and Materials Transactions A. 2005;**36**(4):881-893

[6] Varalakshmi S, Kamaraj M, Murty B. Formation and stability of equiatomic and nonequiatomic nanocrystalline CuNiCoZnAlTi high-entropy alloys by mechanical alloying. Metallurgical and Materials Transactions A. 2010;**41**(10):2703-2709

[7] Zhang K et al. Annealing on the structure and properties evolution of the CoCrFeNiCuAl high-entropy alloy. Journal of Alloys and Compounds. 2010;**502**(2):295-299

[8] Yeh JW et al. High-entropy alloys—A new era of exploitation. In: Materials Science Forum. Taiwan: Trans Tech Publ.; 2007. pp. 1-9

[9] Zhang Y et al. Microstructures and properties of high-entropy alloys. Progress in Materials Science. 2014;**61**:1-93

[10] Gao MC et al. High-Entropy Alloys. United States, China: Springer; 2016. pp. 1-19

[11] Cantor B et al. Microstructural development in equiatomic multicomponent alloys. Materials Science and Engineering A. 2004;**375**:213-218

[12] Yeh J-W. Physical metallurgy of high-entropy alloys. JOM. 2015;**67**(10):2254-2261

[13] Gorsse S, Tancret F. Current and emerging practices of CALPHAD toward the development of high entropy alloys and complex concentrated alloys. Journal of Materials Research. 2018;**33**:1-25

[14] Otto F et al. Relative effects of enthalpy and entropy on the phase stability of equiatomic high-entropy alloys. Acta Materialia. 2013;**61**(7):2628-2638

[15] Miracle DB. High-entropy alloys: A current evaluation of founding ideas and core effects and exploring "nonlinear alloys". JOM. 2017;**69**(11):2130-2136

[16] Kumar A, Rajimwale A, Chopkar M. Phase evolution criteria for AlCoCrCuFeMnSix (x = 0, 0.3, 0.6 and 0.9) high entropy alloys based on experiment and thermodynamic calculation. IOP Conference Series: Materials Science and Engineering. 2018;**383**(1):2-6

[17] Sheng G, Liu CT. Phase stability in high entropy alloys: Formation of

solid-solution phase or amorphous phase. Progress in Natural Science: Materials International. 2011;**21**(6):433-446

[18] Davis JR, Mills K, Lampman S. Metals Handbook. Vol. 1. Properties and Selection: Irons, Steels, and High-Performance Alloys. Materials Park, Ohio 44073, USA: ASM International; 1990. p. 1063

[19] Gwalani B. Developing Precipitation Hardenable High Entropy Alloys. Texas: 2017. pp. 25-30

[20] Dąbrowa J et al. Synthesis and microstructure of the (Co, Cr, Fe, Mn, Ni) $3O_4$ high entropy oxide characterized by spinel structure. Materials Letters. 2018;**216**:32-36

[21] Waterman NA, Dickens P. Rapid product development in the USA, Europe and Japan. World Class Design to Manufacture. 1994;**1**(3):27-36

[22] Excell J, Nathan S. The rise of additive manufacturing. The Engineer. January 2019

[23] Griffith ML et al. Understanding the microstructure and properties of components fabricated by laser engineered net shaping (LENS). MRS Online Proceedings Library Archive. 2000;**625**:1-3

[24] Gokuldoss Prashanth K, Scudino S, Eckert J. Tensile properties of Al-12Si fabricated via selective laser melting (SLM) at different temperatures. Technologies. 2016;**4**(4):38

[25] Beaman JJ et al. Solid Freeform Fabrication: A New Direction in Manufacturing. Norwell, MA: Kluwer Academic Publishers; 1997. pp. 25-49

[26] Herzog D et al. Additive manufacturing of metals. Acta Materialia. 2016;**117**:371-392

[27] Schwab H et al. Selective laser melting of Ti-45Nb alloy. Metals. 2015;**5**(2):686-694

[28] Laakso P et al. Optimization and simulation of SLM process for high density H13 tool steel parts. Physics Procedia. 2016;**83**:26-35

[29] Suryawanshi J et al. Simultaneous enhancements of strength and toughness in an Al-12Si alloy synthesized using selective laser melting. Acta Materialia. 2016;**115**:285-294

[30] Prashanth K, Scudino K, Eckert K. Defining the tensile properties of Al-12Si parts produced by selective laser melting. Acta Materialia. 2017;**126**:25-35

[31] Xiang S et al. Microstructures and mechanical properties of CrMnFeCoNi high entropy alloys fabricated using laser metal deposition technique. Journal of Alloys and Compounds. 2019;**773**:387-392

[32] Zhang MN et al. Microstructure and mechanical behavior of AlCoCuFeNi high-entropy alloy fabricated by selective laser melting. In: Beijing: SFF; 2017. pp. 729-730

[33] Popoola P et al. Laser engineering net shaping method in the area of development of functionally graded materials (FGMs) for aero engine applications—A review. In: Fiber Laser. South Africa: InTech; 2016. pp. 393-394

[34] Zhou P et al. $Al_{0.5}$FeCoCrNi high entropy alloy prepared by selective laser melting with gas-atomized pre-alloy powders. Materials Science and Engineering A. 2019;**739**:86-89

[35] Hagedorn-Hansen D et al. The effects of selective laser melting scan strategies on deviation of hybrid parts. South African Journal of Industrial Engineering. 2017;**28**(3):200-212

[36] Bo Q et al. The helix scan strategy applied to the selective laser melting. The International Journal of Advanced Manufacturing Technology. 2012;**63**(5-8):631-640

[37] Yasa E, Deckers J, Craeghs T, Mohsen B, Kruth JP. Investigation on occurrence of elevated edges in selective laser melting. In: International Solid Freeform Fabrication Symposium; Austin, TX, USA. 2009: 187

[38] Morgan R et al. High density net shape components by direct laser re-melting of single-phase powders. Journal of Materials Science. 2002;**37**(15):3093-3100

[39] Yuan P, Gu D. Molten pool behaviour and its physical mechanism during selective laser melting of TiC/AlSi10Mg nanocomposites: Simulation and experiments. Journal of Physics D: Applied Physics. 2015;**48**(3):035303

[40] Zaeh MF, Branner G. Investigations on residual stresses and deformations in selective laser melting. Production Engineering. 2010;**4**(1):35-45

[41] Saarimäki J. Cracks in Superalloys. Vol. 1897. Sweden: Linköping University Electronic Press; 2018. pp. 12-14

[42] Wang W-R, Wang W-L, Yeh J-W. Phases, microstructure and mechanical properties of AlxCoCrFeNi high-entropy alloys at elevated temperatures. Journal of Alloys and Compounds. 2014;**589**:143-152

[43] Tsao T-K et al. High temperature oxidation and corrosion properties of high entropy superalloys. Entropy. 2016;**18**(2):62

[44] Hemphill MA et al. Fatigue behavior of $Al_{0.5}CoCrCuFeNi$ high entropy alloys. Acta Materialia. 2012;**60**(16):5723-5734

Chapter 8

Flight Vehicle Performance

Aram Baghiyan

Abstract

In this chapter, the problem of flight vehicle performance is described. Performance parameters, such as lift-to-drag ratio, maximum and minimum level flight speed, speeds for the best rate of climb, steepest climb, maximum range and endurance, and most economical climb are described using graphical methods, such as drag polar, Zhukovsky curves with combination of analytical derivations. The approach of graphical description of flight vehicle performance allows to understand the physical basics of the aerodynamic properties of flight vehicles easier and to develop deeper connectivity between their interpretations. In addition, flight envelope and operational limits are discussed using both analytical and graphical methods for better understanding.

Keywords: aerodynamics, flight vehicle, drag polar, flight performance, Zhukovsky curves

1. Introduction

During the flight, aircraft is usually loaded by four forces (**Figure 1**)—gravity force G, lift L, drag D, and thrust T. The combination of these forces defines the behavior of flight and the performance of aircraft. There are also cases when the side forces S act on aircraft due to sideslip (**Figure 1**), which are balanced by the vertical tail of aircraft, and during the analysis of the flight vehicle performance they are usually neglected.

As we can see from the **Figure 1**, the vector of aircraft speed $\vec{v}$ does not coincide with its body frame axes x_b, y_b, and z_b and has inclinations from axis x_b expressed by flow angles—angle of attack α and sideslip angle β. The point of acting gravity force is the center of gravity O_{cg} and usually the line of acting of thrust passes through the center of gravity to avoid generation of destabilizing torques or moments. Aerodynamic forces act at the point named center of pressure O_{cp}, which is not a fixed and changes its position depending on the angle of attack and the air speed. Therefore, a more stable point—aerodynamic center O_{ac}—is introduced, where the changes of aerodynamic forces act, so the aerodynamic moments at that point do not change with the changes of angle of attack. However, aerodynamic center can have variations [1] with Mach number $M = v/a$, where v is the magnitude of aircraft speed, a is the local speed of sound.

In problems of flight vehicle performance analysis, usually a simpler model of the balance of forces is used: the side forces S are neglected due to small values of sideslip angle β in steady flight regimes, the point of action of all unneglectable forces assumed the same, at the same time, taking the lines of action of thrust T and drag D forces coinciding (**Figure 2**).

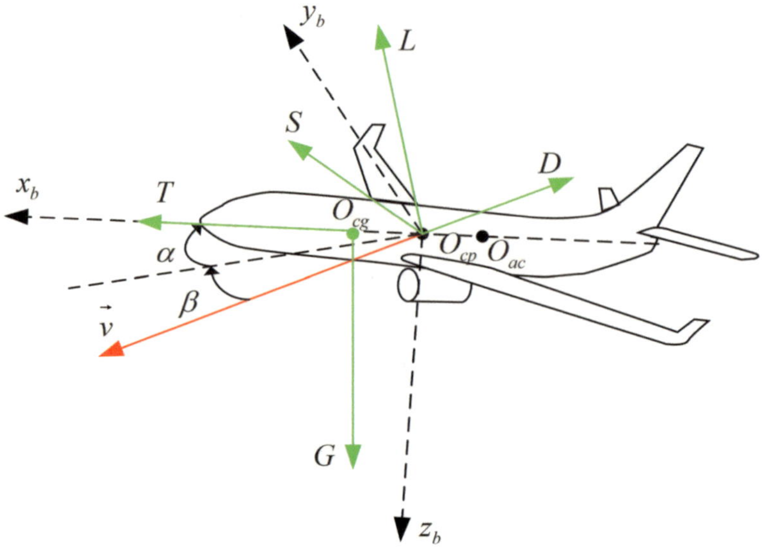

Figure 1.
Forces acting on aircraft.

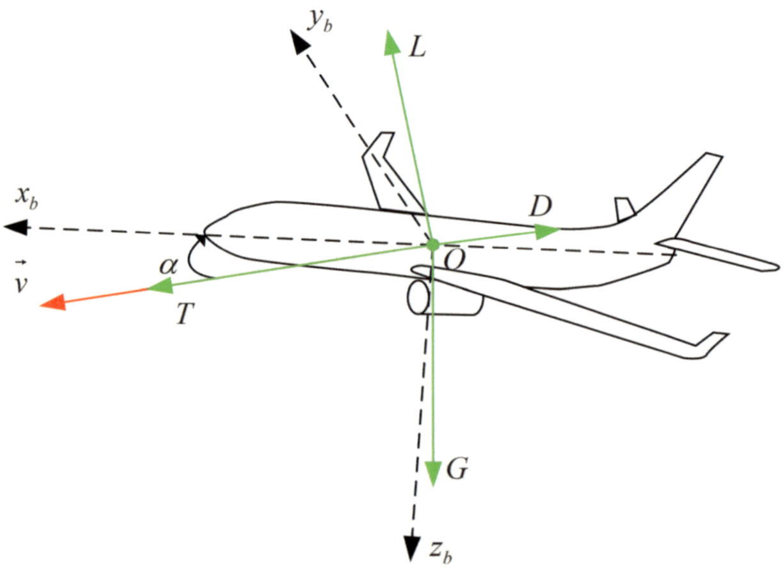

Figure 2.
Simple model of force balance.

Based on the model of force balance, we will study flight vehicle performance at several steady flight regimes, but before getting there let us consider the aerodynamic forces—lift and drag, and the effect of their relations on the aerodynamic quality of aircraft.

2. Drag polar of aircraft

Taking as a reference any paper on aerodynamics, we can find the formulas for lift L and drag D forces [2]:

$$\begin{cases} L = c_L \dfrac{\rho v^2}{2} S_{ref}, \\ D = c_D \dfrac{\rho v^2}{2} S_{ref}, \end{cases} \tag{1}$$

where c_L is the dimensionless lift coefficient; ρ is the freestream density; v is the magnitude of freestream speed, which is taken equal to aircraft speed; S_{ref} is the reference area, and c_D is the dimensionless drag coefficient.

From the system of Eq. (1)

$$\frac{L}{D} = \frac{c_L}{c_D}.$$

The ratio $k = L/D$ is usually called lift-to-drag ratio and in eastern literature it is defined as aerodynamic quality of aircraft. This ratio has interesting properties and it changes with changes of angle of attack α.

Let us consider now the lift coefficient c_L and its relations with angle of attack α (**Figure 3**), which is similar to the analogous relation for 2-D airfoils. The function $c_L = f_L(\alpha)$ can be found using experiments with aircraft model in wind tunnels or using methods of computational fluid dynamics (CFD).

As we can see from the graph of function $c_L = f_L(\alpha)$ (**Figure 3**) there is an angle of attack α_{cr} at which the lift coefficient is maximal, that is, $c_{L\,\max}$. The flight at critical angle of attack α_{cr} will lead to stall, resulting aircraft crash. At angle of attack α_{ts}, the tip stall processes are started and the rate of increase of lift coefficient is decelerated, allowing it to get its maximum value $c_{L\,\max}$ and decrease sufficiently. At the point (α_{ts}, c_{ts}) of starting tip stall, the effects of shaking of aircraft are started [3]. The range between angles of attack α_0, which is called zero lift angle of attack,

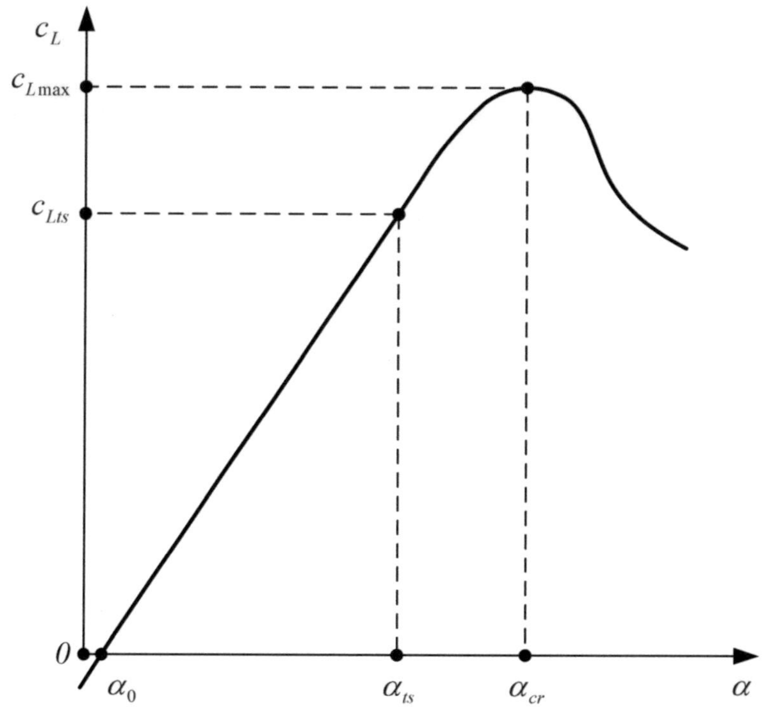

Figure 3.
Lift coefficient of aircraft versus angle of attack.

and α_{ts} is the range of regular flights of aircraft and as it can be observed at this range the function $c_L = f_L(\alpha)$ is linear.

Before getting to the drag coefficient study, let us consider the existing types of the drag force. The drag force D can be represented as a sum of parasitic drag D_p, which consists of form drag D_f and skin friction drag D_{sf} [4], lift-induced D_i drag, wave drag D_w, and interference drag D_{if}:

$$D = D_f + D_{sf} + D_i + D_w + D_{if},$$

or, which is the same as

$$c_D \frac{\rho v^2}{2} S_{ref} = \left(c_{Dp} + c_{Di} + c_{Dw} + c_{Dif}\right) \frac{\rho v^2}{2} S_{ref},$$

where $c_{Dp} = c_{Df} + c_{Dsf}$ is the parasitic drag coefficient, c_{Df} is the form drag coefficient, c_{Dsf} is the skin friction drag coefficient, c_{Di} is the lift-induced drag coefficient, c_{Dw} is the wave drag coefficient, and c_{Dif} is the inference drag coefficient.

Parasitic drag is the pressure difference in front of and behind the wing. The pressure difference depends on the shape of the wing airfoil, its relative thickness $\bar{c}$ and curvature. The larger the relative thickness of the wing airfoil, the greater the form drag (also known as pressure drag); on the other hand, the lower the relative thickness of the wing airfoil, the greater the effect of skin friction drag [5] (**Figure 4**).

The lift-induced drag is the result of the flow tilt (**Figure 5**). Due to the pressure difference above and under the wing on its tips, vortices are generated, leading to the downwash of air from upper surface with velocity u. Thus the effective flow speed v_{eff} becomes the vector sum of the freestream air speed v and downwash speed u. The direction of the effective flow speed differs from the freestream velocity's direction by angle $\delta\alpha$, so the effective angle of attack α_{eff} is defined as:

$$\alpha_{eff} = \alpha + \delta\alpha.$$

With an increase in the angle of attack or lift coefficient, the pressure difference under and above the wing increases quickly, and the coefficient of lift-induced drag increases according to the quadratic law [2]:

$$c_{Di} = \frac{c_L^2}{\pi \lambda e}, \tag{2}$$

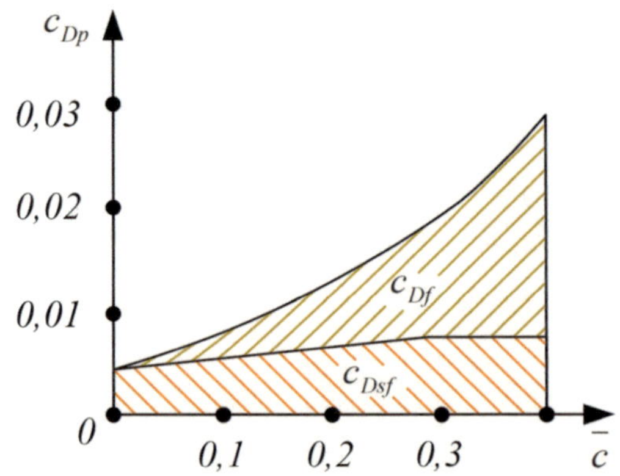

Figure 4.
Parasitic drag components.

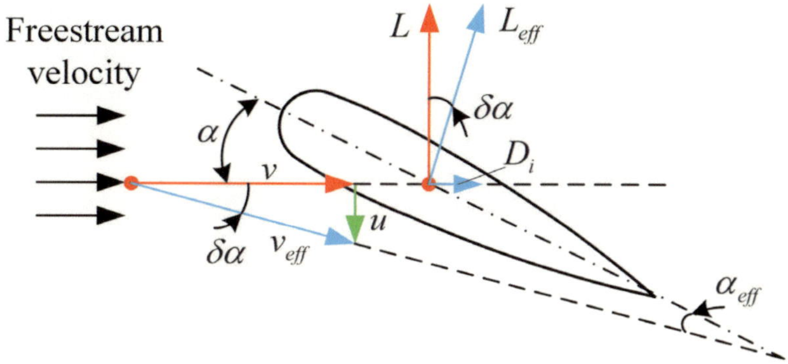

Figure 5.
On lift-induced drag.

where $\lambda = b^2/S_{ref}$ is the aspect ratio, b is the wing span, S_{ref} is the wing reference area, and e is the span efficiency.

The wave drag D_w is a consequence of the compressibility of the air and occurs when there are shock waves near the aircraft.

The mutual influence of the parts of the aircraft is called interference. It occurs due to a change in the velocity field, as a result of which the nature of the flow around the aircraft changes leading to generation of interference drag D_{if}.

Based on the review of drag components, we can divide them into components related to the lift generation or lift-induced drag, and components not related to the lift generation:

$$c_D = c_{D0} + c_{Di}, \tag{3}$$

where c_{D0} is the component of drag coefficient not related with the lift generation and is called zero lift drag coefficient. Usually c_{D0} is taken as constant and not related to the angle of attack, while c_{Di} is proportional to the square of lift coefficient c_L, which linearly depends on angle of attack α in range of the regular flight regimes. Thus the function $c_D = f_D(\alpha)$ should have the graph of parabolic form (**Figure 6**).

Similar to $c_L = f_L(\alpha)$, function $c_D = f_D(\alpha)$ can be found using experiments with aircraft model in wind tunnels and using CFD tools.

Using estimated or experimental results for $f_L(\alpha)$ and $f_D(\alpha)$ at set of angles of attack, it is possible to draw drag polar of aircraft (**Figure 7**).

Based on the drag polar conditions for the best lift-to-drag ratio, zero lift drag and maximal lift can be found. By drawing tangent 1 to the curve of drag polar from the origin of coordinate frame $c_D c_L$, the best lift-to-drag ratio can be found, which corresponds to the angle of attack α_{bldr}; the tangent 2 to the drag polar parallel to axis c_D defines maximal value of lift coefficient $c_{L\max}$ at critical angle of attack α_{cr}, and tangent 3 to the drag polar parallel to axis c_L defines zero lift drag coefficient c_{D0} at zero lift angle of attack α_0.

The angle of attack α_{bldr} of best lift-to-drag ratio has sufficient role in the flight of aircraft, as the flight with this angle provides the maximum value of the mentioned ratio, also known as aerodynamic quality of aircraft:

$$k_{\max} = \frac{f_L(\alpha_{bldr})}{f_D(\alpha_{bldr})}.$$

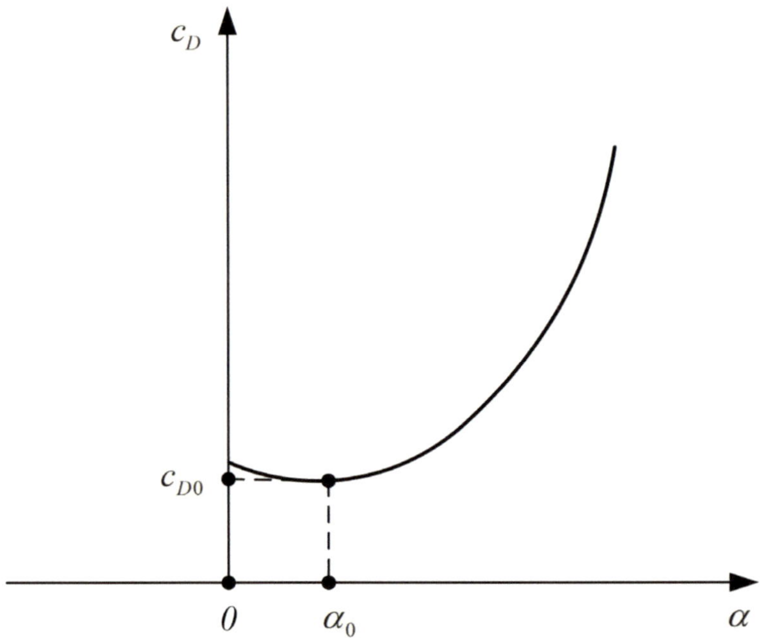

Figure 6.
Drag coefficient of aircraft versus angle of attack.

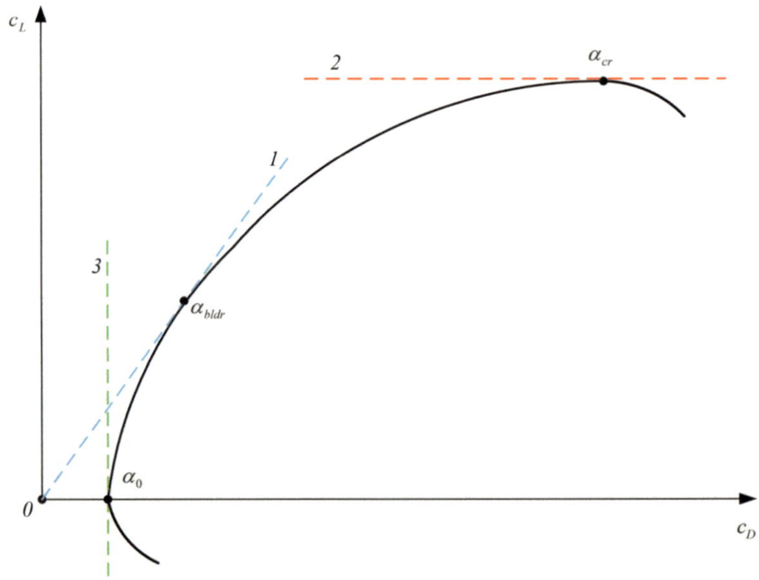

Figure 7.
Drag polar of aircraft.

Let us consider Eq. (3) taking into account Eq. (2):

$$c_D = c_{D0} + \frac{c_L^2}{\pi \lambda e}.$$

The above expression is also called drag polar equation, with the use of which we can represent the non-negative values of lift coefficient as

$$c_L = \sqrt{\pi\lambda e(c_D - c_{D0})}.$$

Taking c_D derivative of c_L we get:

$$\frac{\partial c_L}{\partial c_D} = \partial\frac{\sqrt{\pi\lambda e(c_D - c_{D0})}}{\partial c_D} = \frac{\pi\lambda e}{2\sqrt{\pi\lambda e(c_D - c_{D0})}}.$$

At point $c_D(\alpha_{bldr})$, this derivative is the same as the slope $k_{\max}$ of tangent 1 from **Figure 7**, so we can write down the following expression:

$$\frac{\pi\lambda e}{2\sqrt{\pi\lambda e(c_D(\alpha_{bldr}) - c_{D0})}} = \frac{c_L(\alpha_{bldr})}{c_D(\alpha_{bldr})} = \frac{\sqrt{\pi\lambda e(c_D(\alpha_{bldr}) - c_{D0})}}{c_D(\alpha_{bldr})},$$

which can be easily transformed to

$$c_D(\alpha_{bldr}) = 2c_{D0}. \tag{4}$$

Based on the Eq. (4) we can find the maximal value of lift-to-drag ratio:

$$k_{\max} = \frac{1}{2}\sqrt{\frac{\pi\lambda e}{c_{D0}}}.$$

To examine the dependency of the lift-to-drag ratio on angles of attack, the graph of the function $k(\alpha) = f_L(\alpha)/f_D(\alpha)$ can be plotted (**Figure 8**) and the range of angles near the α_{bldr} studied to find the effective flight regimes and patterns [3, 6].

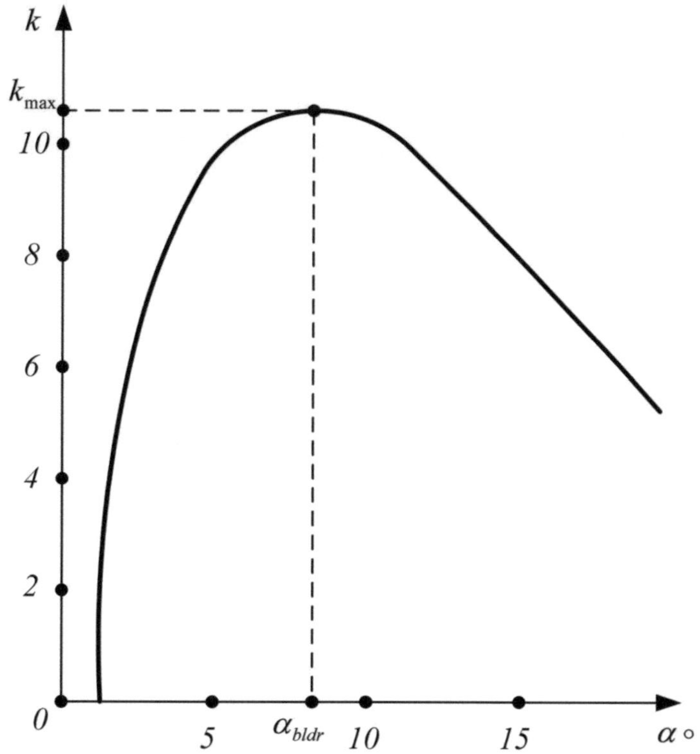

Figure 8.
Lift-to-drag ratio versus angle of attack.

Examined material is one of core bases of aircraft performance, and the results obtained through the above analysis are used in studies of different flight paths and patterns and will be referred in next subsection dedicated to the Zhukovsky curves.

3. Zhukovsky curves

Let us now consider steady horizontal flight. The scheme on **Figure 2** will be transformed to the following form (**Figure 9**):

In steady horizontal flight, we have the following equation of the force balance:

$$\begin{cases} L = G, \\ T = D. \end{cases} \tag{5}$$

which is same as:

$$\begin{cases} c_L \dfrac{\rho v^2}{2} S_{ref} = G, \\ T = c_D \dfrac{\rho v^2}{2} S_{ref} = \left(c_{D0} + \dfrac{c_L^2}{\pi \lambda e} \right) \dfrac{\rho v^2}{2} S_{ref}. \end{cases}$$

From the first equation of the above system, we can find that

$$c_L = \frac{2G}{\rho v^2 S_{ref}},$$

and by substituting the value c_L in the second equation we get:

$$T = c_{D0} \frac{\rho v^2}{2} S_{ref} + \frac{1}{\pi \lambda e} \frac{2G^2}{\rho v^2 S_{ref}}.$$

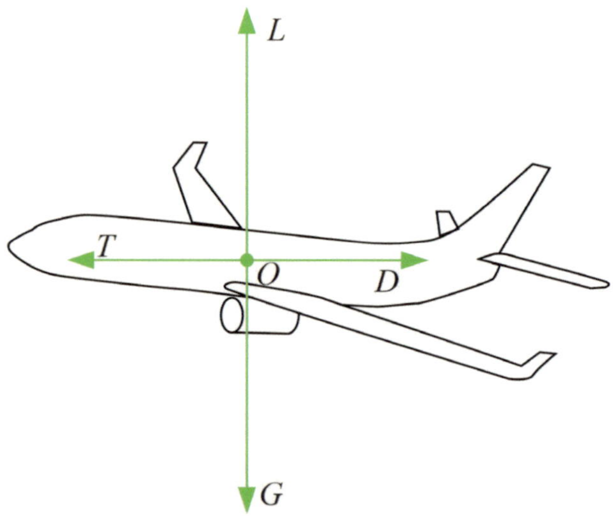

Figure 9.
Force balance at steady horizontal flight.

Based on the above result, we can state that the required thrust T_r for the steady horizontal flight should be equal to the sum of zero lift drag D_0 and lift-induced drag Di, which are defined as:

$$\begin{cases} T_r = D_0 + Di, \\ D_0 = c_{D0}\dfrac{\rho v^2}{2}S_{ref}, \\ Di = c_{Di}\dfrac{\rho v^2}{2}S_{ref} = \dfrac{1}{\pi\lambda e}\dfrac{2G^2}{\rho v^2 S_{ref}}. \end{cases}$$

As we can see, zero lift drag D_0 is proportional to the square of the air speed, while lift-induced drag Di is inversely proportional to the square of air speed.

Let us now define the conditions of minimal drag or, which is the same as, minimal required thrust at steady horizontal flight:

$$\frac{\partial T_r}{\partial v} = 2c_{D0}\frac{\rho v}{2}S_{ref} - 2\frac{1}{\pi\lambda e}\frac{2G^2}{\rho v^3 S_{ref}} = 0.$$

The above expression can be rewritten as:

$$c_{D0}\frac{\rho v^2}{2}S_{ref} = \frac{1}{\pi\lambda e}\frac{2G^2}{\rho v^2 S_{ref}}. \tag{6}$$

It is obvious that the left-hand side of the above is zero lift drag D_0 and the right-hand side is lift-induced drag Di:

$$D_0 = Di,$$

or

$$c_{D0} = c_{Di}.$$

Thus, the drag coefficient is equal:

$$c_D = c_{D0} + c_{Di} = 2c_{D0}.$$

As we remember from the drag polar, the same condition is true for maximal lift-to-drag ratio, so the conditions for minimal required thrust and maximal lift-to-drag ratio are the same. To complete the calculations of all parameters for minimal required thrust, let us derive the expressions for lift coefficient and air speed:

$$c_{Di} = c_{D0} = \frac{c_L^2}{\pi\lambda e},$$

so,

$$c_L = \sqrt{\pi\lambda e c_{D0}},$$

and from Eq. (6)

$$v_{(L/D)\min} = \sqrt[4]{\frac{1}{\pi\lambda e c_{D0}}\frac{4G^2}{\rho^2 S_{ref}^2}}.$$

We can also find the characteristics of available thrust T_a provided by manufacturers of engines or the estimations of available thrust from the sources of literature. In [7, 8], the forms of dependency of available thrust on the air speed for several types of engines are presented. Particularly, in [3] we can find available thrust versus air speed at several altitudes for the jet aircraft L-39 (**Figure 10**).

Based on the available information, Nikolay Zhukovsky developed a graphical method for the analysis of the range of the horizontal flight speeds at different altitudes. His method is based on the plotting curves of zero lift drag and lift-induced drag versus air speed at different altitudes, graphically calculating their sum, and plotting the dependencies of available thrust on air speed at corresponding altitudes for graphoanalytic estimation of ranges of available air speeds for horizontal flight at different altitudes. The set of curves obtained through the above-described procedure in memory of him are called Zhukovsky curves. The graph with curves of required and available thrusts or powers at certain altitude is also called performance diagram.

Let us implement the method proposed Zhukovsky for any altitude, for example, do a plot for the altitude of zero meters above sea level (**Figure 11**) and define the speed characteristics of the horizontal flight.

From Zhukovsky curves at sea level altitude, we can find the maximum speed $v_{\max}$ corresponding to the right point of intersection of curves for required thrust $T_r(v)$ and available thrust $T_a(v)$. At speed $v_{(L/D)\min}$ for which zero lift drag becomes equal to lift-induced drag, $D_0\left(v_{(L/D)\min}\right) = Di\left(v_{(L/D)\min}\right)$, we get minimal required thrust $T_{r\min} = 2D_0\left(v_{(L/D)\min}\right) = 2Di\left(v_{(L/D)\min}\right)$. This speed is a very important

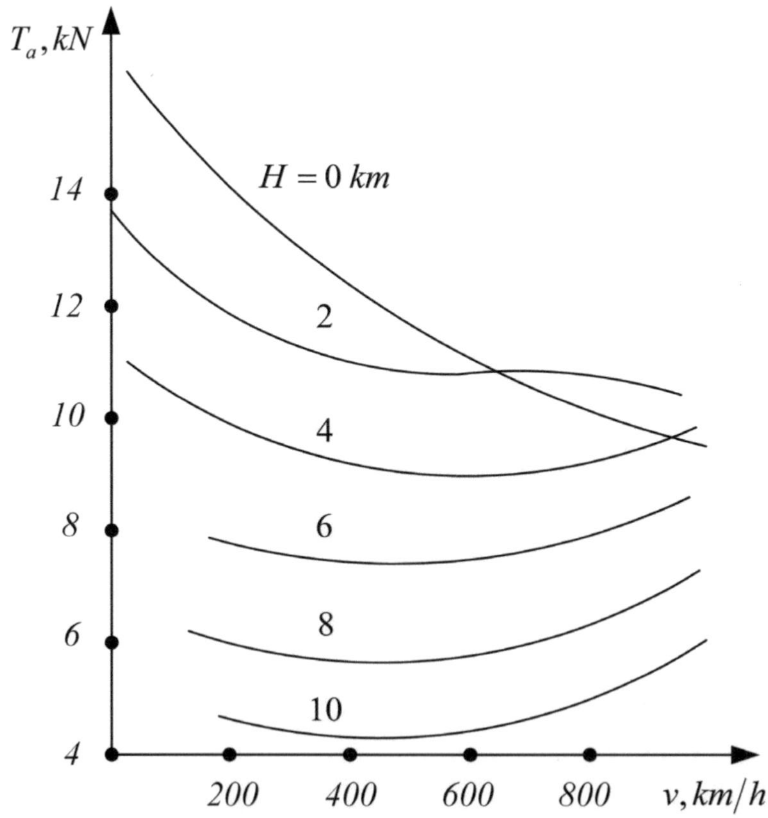

Figure 10.
Available thrust for jet aircraft L-39.

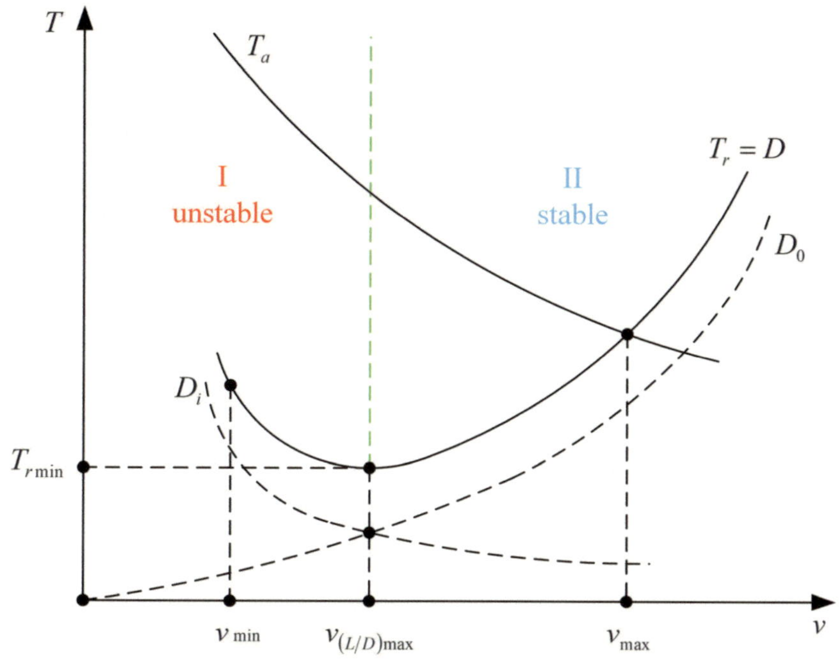

Figure 11.
Zhukovsky curves at sea level altitude.

quantity from the point of horizontal flight, as the flight at lower speeds is not stable to the accidental changes of speed and requires attention of pilots. On the other hand, the flight at higher speeds than $v_{(L/D)\min}$ is stable to accidental changes of speed, as if speed decreases drag force also decreases and as the thrust was at initial value it accelerates the aircraft till there is a balance between thrust and drag. We get a similar picture for accidental increase of speed at stable region II, for which the drag force decelerates the aircraft till drag-thrust balance. For unstable region I, the dangerous case is the accidental decrease of speed, which increases drag leading to decelerating the aircraft till stall. To fly at unstable region of speeds, the pilot needs always to work with the throttle to increase thrust when it is required. Minimum speed $v_{\min}$ at sea level flight is not defined from the above curves and refers to the stall speed, which can be found from the condition of required lift:

$$c_{L\max}\frac{\rho v^2}{2}S_{ref} = G,$$

or

$$v_{\min} = \sqrt{\frac{2G}{c_{L\max}\rho S_{ref}}}.$$

For flight at higher altitudes, we can get conditions when the required thrust at the above minimum speed is much higher than the available thrust at that altitude (**Figure 12**). In such cases, minimum speed defined as minimum thrust speed $v_{T\min}$ corresponds to the left point of intersection of curves for required thrust $T_r(v)$ and available thrust $T_a(v)$.

The idea of plotting Zhukovsky curves at sea level flight allows us to have the same graph at any altitude H for the required trust $T_r(v_{IAS})$ versus indicated

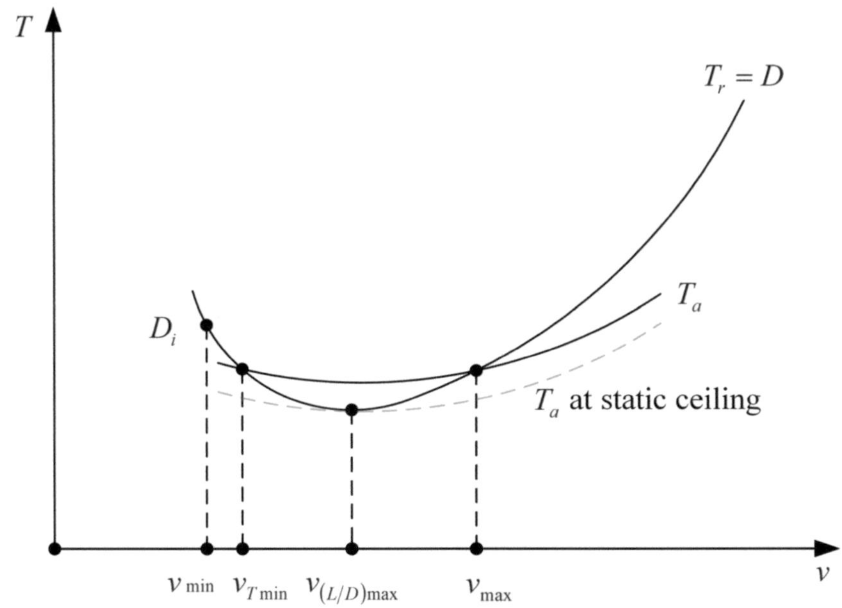

Figure 12.
On minimum thrust speed.

airspeed (IAS), which is usually measured on aircraft and related to the true airspeed (TAS) v_{TAS} via the expression:

$$v_{IAS} = v_{TAS}\sqrt{\frac{\rho_H}{\rho_0}},$$

where ρ_0 is the air density at sea level and ρ_H is the air density at altitude H above sea level (ASL).

Based on the Zhukovsky curves based on IAS, we can define the theoretical or static ceiling of horizontal flight (**Figure 12**), which is an altitude where the horizontal flight is possible only with IAS equal to $v_{(L/D)\,\mathrm{min}}$. The service ceiling has a more practical meaning, as it is the altitude where rate of climb (ROC) becomes less than 0.5 m/s [6, 7].

There are also defined such concepts as speed of maximum endurance v_e, cruise speed v_{cr}, and speed of maximum range v_r. The maximum endurance speed and maximum range speed depend on fuel consumption characteristics of engine. The speed corresponding to the minimum hourly fuel consumption of engine is called speed of maximum endurance or economical speed. On the other hand, the speed corresponding to the minimum per-kilometer consumption of engine is the speed of maximum range and it is very close to the cruise speed (slightly more than cruise speed for real aircraft). The cruise speed is the speed at which the ratio of drag to speed is minimal, and can be found using Zhukovsky curves by drawing a tangent to the required thrust graph from the origin (**Figure 13**).

Let us now consider descending flight or glide (**Figure 14a**) and ascending (**Figure 14b**) flight of aircraft.

In descending flight, the throttle is usually set to minimum, so the thrust can be neglected and consider the gliding flight (**Figure 14a**), for which we can write down the following equations of force balance:

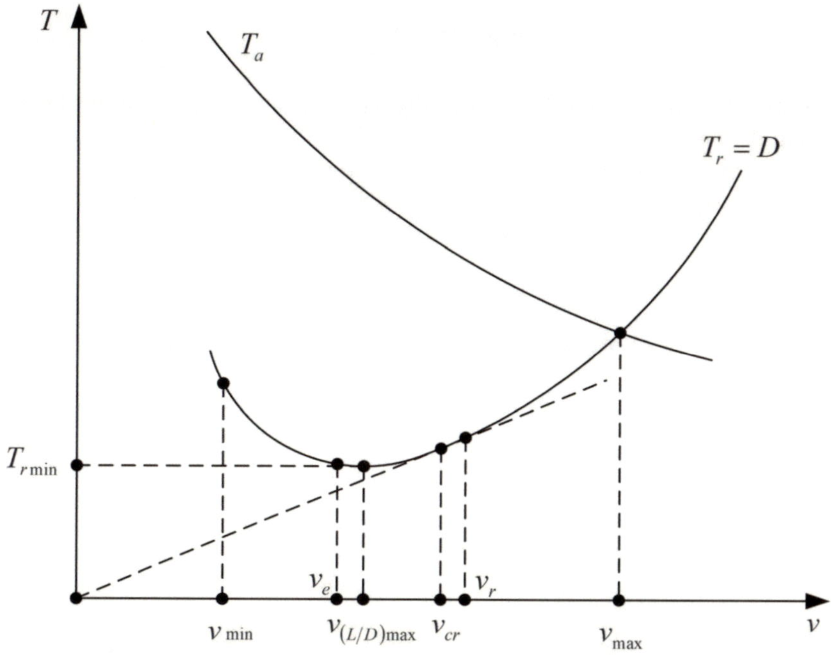

Figure 13.
On cruise and other speeds.

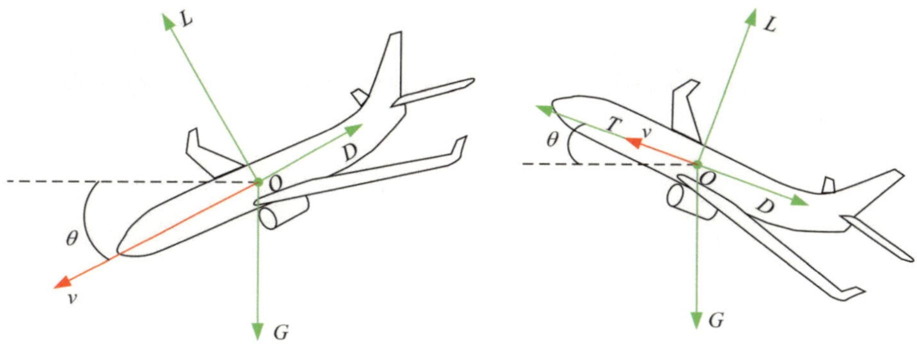

Figure 14.
Force balance for descending and ascending flights. (a) Gliding flight; (b) ascending flight.

$$\begin{cases} L = G\cos|\theta|, \\ D = G\sin|\theta|. \end{cases}$$

where θ is the flight path angle.
Based on the above we can get:

$$\tan|\theta| = \frac{D}{L} = \frac{1}{k}.$$

The flattest glide corresponds to the minimum magnitude of flight path angle $|\theta|_{min}$, which is case when the $\tan|\theta|$ is minimum, thus, we can write down the following:

$$|\theta|_{min} = \operatorname{atan}\frac{1}{k_{\max}}.$$

The above means that the flattest glide, resulting the longest gliding distance, is also related to maximum lift-to-drag ratio $k_{\max}$.

For the ascending flight the force balance is presented as follows:

$$\begin{cases} L = G\cos\theta, \\ T - D = G\sin\theta. \end{cases} \tag{7}$$

As we know, the rate of climb is the projection of airspeed to the vertical plane:

$$ROC = v\sin\theta.$$

From second equation of Eq. (7) we can get:

$$\sin\theta = \frac{T-D}{G},$$

or which is same as:

$$ROC = v\sin\theta = \frac{Tv - Dv}{G} = \frac{P_a - P_r}{G},$$

where $P_a = Tv$ is the available power, $P_r = Dv$ is the required power.

From the above we can find the maximum of ROC, that is, $ROC_{\max}$ corresponding to maximum of excess power $\Delta P_{\max} = (P_a - P_r)_{\max}$ (**Figure 15**).

The angle $\theta_{\max}$ of steepest ascending flight can be found by dividing the ROC value by speed [9]:

$$\theta_{\max} = \operatorname{asin}\left(\frac{ROC}{v}\right)_{\max},$$

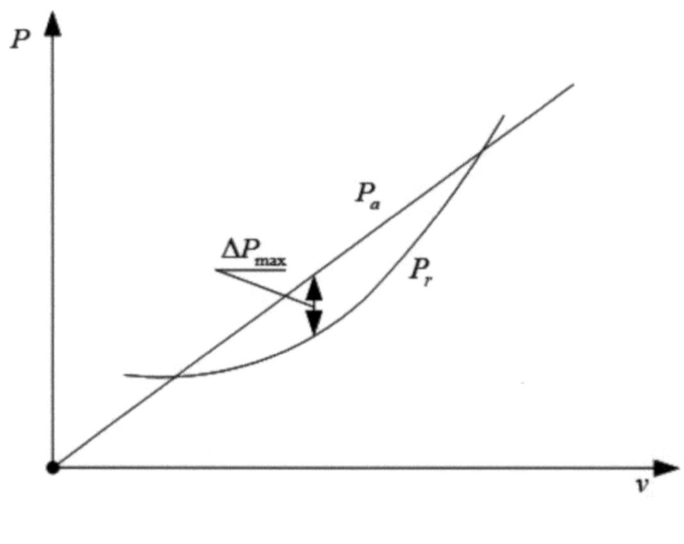

Figure 15.
Power diagram for jet aircraft.

which corresponds to

$$\theta_{\max} = \operatorname{asin}\left(\frac{(T_a - T_r)_{\max}}{G}\right).$$

In ideal jet aircraft case, the angle of steepest climb is defined at the speed of maximum lift-to-drag ratio. The most economical ascending flight is defined by the operational regime of aircraft engine at which the minimum fuel will be consumed for aircraft to climb to the required altitude.

4. Flight envelope and operational limits of aircraft

For any aircraft, performance and operational limits are defined. Performance limits (**Figure 16**) are mostly defined by the aerodynamic configuration of aircraft. On the other hand, operational limits are based on the type of aircraft, structural and engine limits, wind resistance parameters, and maximum Mach number. All these limits are presented in diagrams of altitude h versus airspeed v and the final diagram, which represents the intersection of all limits, is called flight envelope (**Figure 17**). The flight envelope can be represented in graphs of altitude versus true airspeed, altitude versus indicated airspeed, or altitude versus Mach number.

All the speeds presented on **Figure 16** were described in the previous subsection. Operational limits related to the type of aircraft are described from the point of view of aircraft application: if an aircraft is a passenger jet, it should apply to the requirements of comfort for passengers and not exceed load factors of comfortable flight; on the other hand, if an aircraft is a jet fighter its operational limits from the point of view of load factors should be derived from compromise between the prevention of health issues of pilot that may occur and maneuverability for the combat use.

Structural limits are mostly related to aircraft strength, while the engine limits can be the result of its design and performance at higher altitudes. Wind resistance limits can be derived from the requirements of operational use or comfortability of flight for passengers.

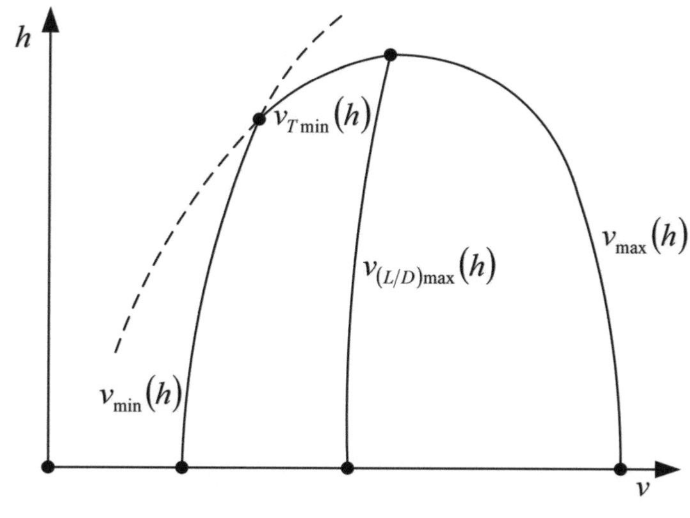

Figure 16.
Performance limits.

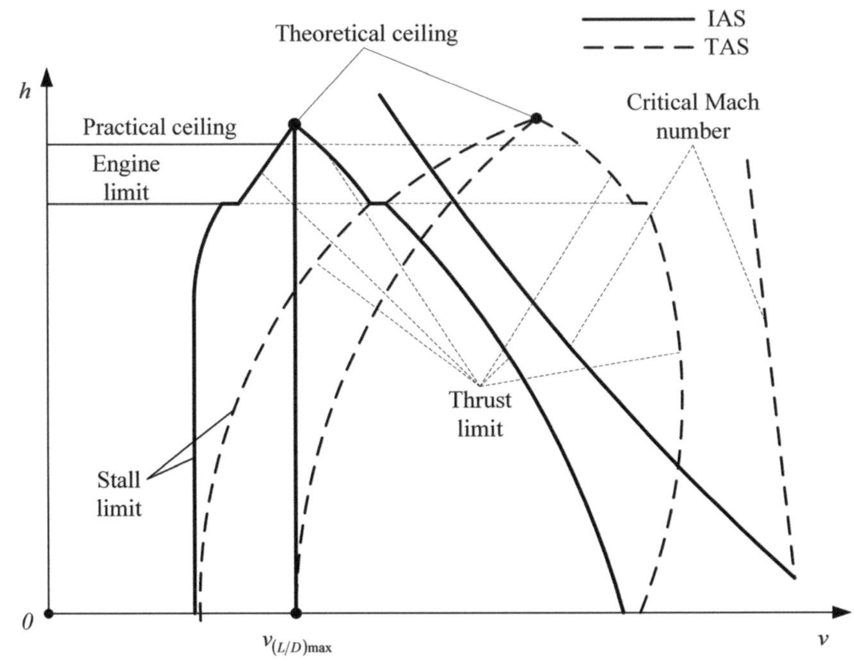

Figure 17.
Example of a flight envelope.

Maximum Mach number can be defined from the conditions of aeroelastic effects and vibrations, effects of shifting aerodynamic center, causing severe pitching moments, which can lead to the crashes, or loosing effectiveness of aerodynamic surfaces. For example, for jet aircraft L-39 the critical Mach number is $M = 0.8$ and exceeding this condition leads to the shift of aerodynamic center of L-39, which generates pitching moment, causing a descending flight with acceleration [3]. To prevent such unstable flight, on L-39 air brakes are used, which automatically act at the Mach numbers $M = 0.78$. The critical Mach number is defined at airspeed flight that leads to generation of shock waves on the wing due to the acceleration of airflow on the upper surface of wing.

Finally, an example of flight envelope for analysis is presented in **Figure 17** where the intersection of all limiting conditions is described.

5. Conclusion

The material described in this chapter involves flight vehicle analysis using graphical and analytical tools for better understanding of the physical aspects of flight core parameters and development of strong and meaningful connections between them. The material from this chapter can be useful in the preliminary design and prototyping of flight vehicles and for finding the paths for further developments and improvements in the design.

Acknowledgments

In this work, the name of Prof. Nikolay Zhukovsky (January 17, 1847–March 17, 1921) is mentioned many times, while his name may be unfamiliar to many readers. Nikolay Zhukovsky was one of the first scientists who established the mathematical base of aerodynamics; he was the initiator and the first head of the Central

Aerohydrodynamic Institute (TsAGI), and on his initiative was created Zhukovsky Air Force Engineering Academy—alma mater of three famous Soviet aircraft designers Sergei Ilyushin, Artem Mikoyan, and Alexander Yakovlev.

Science lives in the research schools, and great scientists are those who both do great discoveries and develop the next generations of discoverers. Zhukovsky Air Force Engineering Academy is my alma mater as well, and with these few words I would like to express my appreciation to all my teachers who helped me on my way of professional and personal development.

Author details

Aram Baghiyan
Improvis Aerospace and Defense LLC, Yerevan, Armenia

*Address all correspondence to: abaghiyan@improvismail.com

References

[1] Lamar JE, Alford WJ Jr. Aerodynamic Center Considerations of Wings and Wing-Body Combinations. NASA Technical Note, TN D-3581; 1966. p. 16

[2] Drela M. Flight Vehicle Aerodynamics. Cambridge, Massachusetts: The MIT Press; 2014. p. 304

[3] Lysenko NM, editor. Practical Aerodynamics of Training Jet Aircraft. P. 2. Practical Aerodynamics of Aircraft L-39. Moscow: Voennoe Izdatelstvo; 1985. p. 248

[4] Clancy LJ. Aerodynamics. New York: Wiley; 1975. p. 172

[5] Korovin AE, Novikov YF. Practical Aerodynamics and Flight Dynamics of the Yak-52 and Yak-55 Aircraft. Moscow: DOSAAF; 1989. p. 357

[6] Kermode AC. In: Barnard RH, Philpott DR, editors. Mechanics of Flight (11th Edition)/Rev. London: Prentice Hall; 2006. p. 512

[7] Anderson JD. Introduction to Flight. New York: McGraw-Hill Education; 2016. p. 928

[8] Sloop JL. Liquid Hydrogen as a Propulsion Fuel. Washington, D.C.: NASA Scientific and Technical Information Office; 1945-1959, 1978. p. 338

[9] Smetana FO. Flight Vehicle Performance and Aerodynamic Control. Reston: American Institute of Aeronautics and Astronautics; 2001. p. 359

Chapter 9

Critical Mach Numbers of Flow around Two-Dimensional and Axisymmetric Bodies

Vladimir Frolov

Abstract

The paper presents the calculated results obtained by the author for critical Mach numbers of the flow around two-dimensional and axisymmetric bodies. Although the previously proposed method was applied by the author for two media, air and water, this chapter is devoted only to air. The main goal of the work is to show the high accuracy of the method. For this purpose, the work presents numerous comparisons with the data of other authors. This method showed acceptable accuracy in comparison with the Dorodnitsyn method of integral relations and other methods. In the method under consideration, the parameters of the compressible flow are calculated from the parameters of the flow of an incompressible fluid up to the Mach number of the incoming flow equal to the critical Mach number. This method does not depend on the means determination parameters of the incompressible flow. The calculation in software Flow Simulation was shown that the viscosity factor does not affect the value critical Mach number. It was found that with an increase in the relative thickness of the body, the value of the critical Mach number decreases. It was also found that the value of the critical Mach number for the two-dimensional case is always less than for the axisymmetric case for bodies with the same cross-section.

Keywords: compressibility, critical Mach number, air, two-dimensional case, axisymmetric case

1. Introduction

This chapter provides an overview of the results obtained by the author by an approximate method for determining the critical Mach numbers for flows in two-dimensional and axisymmetric cases [1–3].

The high subsonic flow velocities in aerodynamics are a common case since most modern passenger and cargo planes fly at Mach numbers exceeding the critical Mach numbers. The compressibility effect increases with an increase in the free-stream Mach number and with an increase in the perturbations created by the flying body at low Mach numbers. Compressibility problems have been considered by many scientists by various methods [4–20]. Knowing the free-stream Mach number, at which the local velocity somewhere on the surface of the body becomes equal to the local sound velocity, makes it possible to correctly choose the basic aerodynamic equations for a body moving at high speed. It is well known that when the local flow velocity becomes equal to the speed of sound, the aerodynamic equations change their own type from elliptic to hyperbolic. The value of the critical Mach number is the transition from one type of equation to another.

2. The approximate method for calculating flow compressibility characteristics

The equations for the existence of the velocity potential and continuity are written for two cases: two-dimensional and axisymmetric for compressible irrotational gas flow

$$\frac{\partial u}{\partial r} - \frac{\partial v}{\partial x} = 0, \quad \frac{\partial (r^n \rho u)}{\partial x} + \frac{\partial (r^n \rho v)}{\partial r} = 0, \tag{1}$$

where u, v are components of the velocity along axis x и r accordingly, *m/s*;, *r, x* are coordinates; ρ is local density of gas, *kg/m*^{3}; parameter n equal

$$n = \begin{cases} 0 & - \text{2D body;} \\ 1 & - \text{axisymmetric body.} \end{cases}$$

Here x, r are coordinates in the meridional plane for the axisymmetric case (n = 1). Let us introduce the special functions proposed by Burago [20]

$$\tau = \frac{2}{\rho_0/\rho + 1}; \quad \sigma = \frac{\rho_0/\rho - 1}{\rho_0/\rho + 1}, \tag{2}$$

ρ_0 is gas density at the stagnation point. Taking into account equation (Eq. (2)), the equation (Eq. (1)) can be rewritten as

$$\frac{\partial}{\partial r}\left(\frac{\tau u}{1-\sigma}\right) - \frac{\partial}{\partial x}\left(\frac{\tau u}{1-\sigma}\right) = 0, \quad \frac{\partial}{\partial x}\left(\frac{r^n \tau u}{1+\sigma}\right) + \frac{\partial}{\partial r}\left(\frac{r^n \tau v}{1+\sigma}\right) = 0, \tag{3}$$

For barotropic model of compressible air we have

$$\frac{\rho}{\rho_\infty} = \left(\frac{p}{p_\infty}\right)^{\frac{1}{\kappa}}, \tag{4}$$

where p is static pressure, *Pa*; κ is ratio of specific heats (for air κ = 1.4); ∞ subscript indicates flow parameters at infinity. The equation (Eq. (4)) refers to as Poisson's adiabatic curve.

According to the accepted model of a barotropic gas, the enthalpy and pressure function differ only by a constant value $h = P(p) + \text{const}$. The pressure function is

$$P(p) = \int_{p_\infty}^{p} \frac{dp}{\rho}. \tag{5}$$

If we carry out the integration in formula (Eq. (5)) taking into account formula (Eq. (4)), then we get (Eq. (6))

$$h = \frac{\rho^{\kappa-1}}{(\kappa-1)\rho_\infty^{\kappa-1} M_\infty^2}, \tag{6}$$

here M_∞ is Mach number at infinity.

Using the equation for enthalpy (Eq. (6)), the Bernoulli equation for a compressible gas can be written as

$$\frac{|\mathbf{V}|^2}{2} + \frac{\rho^{\kappa-1}}{(\kappa-1)\rho_\infty^{\kappa-1}M_\infty^2} = \frac{1}{2} + \frac{1}{(\kappa-1)M_\infty^2}, \tag{7}$$

where $|\mathbf{V}|$ is the modulus of the vector of the total local velocity of flow.

Based on formula (Eq. (7)), the following equalities can be written:

$$\frac{\rho}{\rho_\infty} = \left[1 + \frac{\kappa-1}{2}M_{\infty 2}\left(1 - \overline{u}^2 - \overline{v}^2\right)\right]^{\frac{1}{\kappa-1}}, \quad \frac{a}{a_\infty} = \left[1 + \frac{\kappa-1}{2}M_{\infty 2}\left(1 - \overline{u}^2 - \overline{v}^2\right)\right]^{\frac{1}{2}}, \tag{8}$$

here $\overline{u} = u/U_\infty$, $\overline{v} = v/U_\infty$ are the dimensionless components of the local velocity, U_∞ is free-stream flow velocity at infinity.

Let us write the well-known isentropic relations for density and sound speed

$$\frac{\rho_0}{\rho} = \left[1 + \frac{\kappa-1}{2}M^2\right]^{\frac{1}{\kappa-1}}, \quad \frac{a_0}{a} = \left[1 + \frac{\kappa-1}{2}M^2\right]^{\frac{1}{2}}, \tag{9}$$

here lower index "0" specifies parameters in stagnation point.

Formulas (9) can be written as

$$\frac{\rho_0}{\rho} = [E(M)]^{\frac{1}{\kappa-1}}, \quad \frac{a_0}{a} = [E(M)]^{\frac{1}{2}}, \quad E(M) = 1 + \frac{\kappa-1}{2}M^2. \tag{10}$$

The speed of sound is defined by formula $a = \sqrt{dp/d\rho}$. The speed of sound in the case of an isentropic air flow is $a = \sqrt{\kappa p/\rho}$. Based on the equation of state in the form (Eq. (4)), we can write

$$a = a_\infty \left(\frac{\rho}{\rho_\infty}\right)^{\frac{\kappa-1}{2}}. \tag{11}$$

Based on the isentropic relationship for density (Eq. (9)), the function σ (Eq. (2)) can be calculated. **Figure 1** shows a graph of the function σ for air versus Mach number.

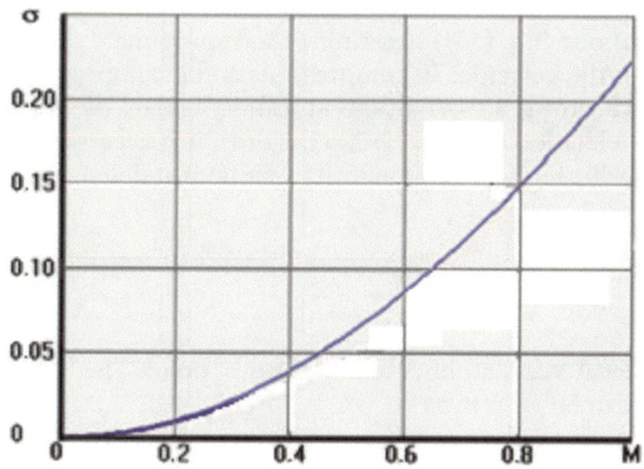

Figure 1.
The function σ vs. Mach number.

In **Figure 1** it is seen that the values σ lie in a narrow range [0; 0.22] for air when changing the Mach number from 0 to 1.0. This narrow range allows us to take $\sigma \approx$ const, and then using formulas (Eq. (2)), we can replace (Eq. (3)) with an approximate

$$\frac{\partial}{\partial x}\left(\frac{r^n \tau u}{\rho_0 + \rho}\right) + \frac{\partial}{\partial r}\left(\frac{r^n \tau v}{\rho_0 + \rho}\right) = 0, \quad \frac{\partial}{\partial r}\left(\frac{\tau u}{\rho_0 + \rho}\right) - \frac{\partial}{\partial x}\left(\frac{\tau u}{\rho_0 + \rho}\right) = 0. \tag{12}$$

If we introduce new variables u° and v° associated with u and v as follows

$$u = \frac{\rho_\infty}{\rho}\frac{(\rho + \rho_0)}{(\rho_\infty + \rho_0)} u^0 = \frac{1 + \frac{\rho_0}{\rho}}{1 + \frac{\rho_0}{\rho_\infty}} u^0 = \eta_c u^0; \quad v = \frac{\rho_\infty}{\rho}\frac{(\rho + \rho_0)}{(\rho_\infty + \rho_0)} v^0 = \frac{1 + \frac{\rho_0}{\rho}}{1 + \frac{\rho_0}{\rho_\infty}} v^0 = \eta_c v^0. \tag{13}$$

where η_c is introduced

$$\eta_c = \left(1 + \frac{\rho_0}{\rho}\right) \Big/ \left(1 + \frac{\rho_0}{\rho_\infty}\right). \tag{14}$$

The coefficient η_c will be called the coefficient of compressibility. The new speeds (Eq. (13)) allow equations (Eq. (12)) to be written in the form

$$\frac{\partial u^0}{\partial r} - \frac{\partial v^0}{\partial x} = 0, \quad \frac{\partial (r^n u^0)}{\partial x} + \frac{\partial (r^n v^0)}{\partial r} = 0. \tag{15}$$

Equations (Eq. (15)) repeat equations (Eq. (1)) describing potential flow of an incompressible liquid ($\rho \approx const$) with a velocity having components (u^0, v^0). Equations (Eq. (15)) allow us to assert that the boundary conditions at infinity for a compressible gas flow with velocity components (u, v) will be identical to the corresponding conditions for an incompressible fluid flow with velocity components (u^0, v^0). The boundary conditions on the body surface will also be the same. Indeed, in a compressible flow on the surface of a body $v_n = 0$. On the body surface in an incompressible fluid flow, based on (Eq. (13)), we can write

$$v_n^0 = \frac{\rho}{\rho_\infty}\frac{(\rho_\infty + \rho_0)}{(\rho + \rho_0)} v_n = 0. \tag{16}$$

Thus, the equations (Eq. (13)) determine the approximate relationship between the components of the velocities in compressible and incompressible flows around the same body under the same conditions at infinity and on the surface of the body. So, to determine velocities of compressible gas flow, it is necessary to use the known isentropic relationships for the local velocity of a flow and density

$$\frac{|\mathbf{V}|}{a_0} = \frac{M}{\sqrt{1 + \frac{\kappa - 1}{2} M^2}}, \quad \frac{\rho_0}{\rho_\infty} = \left[1 + \frac{\kappa - 1}{2} M_\infty^2\right]^{\frac{1}{\kappa - 1}}, \tag{17}$$

here M is the local Mach number at an arbitrary point. The left side of the first formula (Eq. 17) can be written as

$$\frac{|\mathbf{V}|}{a_0} = \frac{U_\infty a_\infty \sqrt{\overline{u}^2 + \overline{v}^2}}{a_\infty a_0} = \frac{M_\infty \sqrt{\overline{u}^2 + \overline{v}^2}}{\sqrt{1 + \frac{\kappa - 1}{2} M_\infty^2}}. \tag{18}$$

From (Eq. (17)) and (Eq. (18)) it follows

$$M = M_\infty \sqrt{\frac{\overline{u}^2 + \overline{v}^2}{1 + \frac{\kappa - 1}{2} M_\infty^2 \left(1 - \overline{u}^2 - \overline{v}^2\right)}}. \tag{19}$$

Equation (Eq. 19) shows that the local Mach number M is determined by the local dimensionless velocities $\overline{u}$ and $\overline{v}$, which, according equations (Eq. 13) and (Eq. 10) depends on local Mach number. To account for this, we will use successive approximations. We will use the velocities calculated for an incompressible flow, which we will substitute in (Eq. (19)) at the first stage of approximation

$$\overline{u}^{(1)} = u^0 / U_\infty; \quad \overline{v}^{(1)} = v^0 / U_\infty. \tag{20}$$

Thus, based on these values, the local Mach number $M^{(1)}$ for the first approximation is determined. Thereafter, the computed first approximation Mach number $M^{(1)}$ is used to calculate the second approximation velocity components from $\overline{u}^{(2)}$, $\overline{v}^{(2)}$ from (Eqs. (13) and (14)), and isentropic formula for density (Eq. (10)). The obtained values of the velocity components make it possible to calculate on the basis (Eq. (19)), the local Mach number for the second approximation $M^{(2)}$, which, in turn, is used to calculate the local velocity components at the stage of the third approximation $\overline{u}^{(3)}, \overline{v}^{(3)}$, and so on. The approximation process must be continued until the following inequality is satisfied $\left|\eta_c^{(n)} - \eta_c^{(n-1)}\right| - \left|\eta_c^{(n-1)} - \eta_c^{(n-2)}\right| \le \varepsilon$, where ε is a given calculation error. Preliminary calculations using the described algorithm showed that the number of approximation steps increases as $M \to 1.0$, but usually the maximum number of approximations does not exceed 30.

From (Eq. (18) and (19)) we can write

$$\frac{M_\infty \sqrt{\overline{u}^2 + \overline{v}^2}}{\sqrt{1 + \frac{\kappa - 1}{2} M_\infty^2}} = \frac{M}{\sqrt{1 + \frac{\kappa - 1}{2} M^2}}. \tag{21}$$

Putting in the formula (Eq. (21)) $M = 1$ and $M_\infty = M_*$, we can get the formula for calculating the critical Mach number

$$M_* = \sqrt{\frac{1}{\overline{u}^2 + \overline{v}^2 + \frac{\kappa - 1}{2}\left(\overline{u}^2 + \overline{v}^2 - 1\right)}} \equiv \sqrt{\frac{1}{\overline{U}^2 + \frac{\kappa - 1}{2}\left(\overline{U}^2 - 1\right)}}, \tag{22}$$

where the notation $\overline{U}$ is introduced for the local total relative velocity. From (Eq. (22)) it can be seen that the minimal value M_* corresponds to the maximal value $\overline{U}$. Therefore, to calculate the critical Mach number for a flow near a body, it is necessary to determine the maximum local velocity on the body surface. The dimensionless velocities $\overline{u}$ and $\overline{v}$ (Eq. (22)) also should correspond to the critical free-stream Mach number. For this, it is necessary to apply a method of successive approximations, similar to that used to calculate the local Mach number (Eq. (19)). As a first approximation, we will use the velocities $\overline{u}^{(1)}$ and $\overline{v}^{(1)}$ calculated for an incompressible flow. The critical Mach number $M_*^{(1)}$ attained in the first approximation should be considered as the free-stream Mach number for calculating the local relative velocities for the second approximation $\overline{u}^{(2)}$ and $\overline{v}^{(2)}$. The approximation process must be continued until the calculated critical Mach number for subsequent approximations changes by a given error ε, i.e., until the condition is satisfied $\left|M_*^{(i)} - M_*^{(i-1)}\right| \le \varepsilon$. It turns out that the critical Mach number can be calculated using only the values of the incompressible flow speeds. It should be noted that the method described above is applicable for compressible flows for which the inequality $M_\infty \le M_*$ is valid. This condition determines the absence of transonic and supersonic zones in the flow field. To calculate the critical Mach number M_* and the velocity field (u, v) in a compressible flow, it is sufficient to calculate the velocity field (u^0, v^0) around the body for an incompressible flow.

It follows from the described method that the critical Mach number for a given 2-D body can be calculated using only the values of the incompressible flow speeds. The method is applicable for compressible flows for which the free-stream Mach numbers are less than the critical Mach number. So, to calculate the critical Mach number, for an incompressible flow, it is sufficient to calculate the velocity field near the streamlined body, but in practical aerodynamics, the pressure distribution on the body surface is obtained experimentally more often than the velocity field. To compare the results of calculations and experimental data, it is convenient to use the pressure-drop coefficient. This circumstance leads to the necessity of obtaining a connection between the pressure field in a compressible flow and the pressure field in an incompressible flow. To establish this connection we write down the pressure-drop coefficient for an incompressible flow

$$c_p^0 = \frac{p^0 - p_\infty}{0,5\rho V_\infty^2} = 1 - \left(\frac{V^0}{V_\infty}\right)^2. \tag{23}$$

Based on formulas (Eq. (13)), the pressure-drop coefficient for an incompressible flow can be represented through the gas flow velocity

$$c_p^0 = 1 - \left(\frac{V}{\eta_c V_\infty}\right)^2. \tag{24}$$

The pressure-drop coefficient for the compressible gas flow can be written

$$c_p = \frac{p - p_\infty}{0.5\rho_\infty V_\infty^2} = 2\frac{p_\infty}{\rho_\infty V_\infty^2}\left(\frac{p}{p_\infty} - 1\right) \tag{25}$$

Let us rewrite the last formula in terms of the Mach number of the free-stream flow. Using the barotropic model of compressed air, we can write the following equalities

$$\frac{p_\infty}{\rho_\infty V_\infty^2} = \frac{1}{\kappa}\frac{a_\infty^2}{V_\infty^2} = \frac{1}{\kappa M_\infty^2},$$

then we rewrite formula (Eq. (25)) in the form

$$c_p = \frac{2}{\kappa M_\infty^2}\left(\frac{p}{p_\infty} - 1\right) = \frac{2}{\kappa M_\infty^2}\left(\frac{p}{p_0}\frac{p_0}{p_\infty} - 1\right). \tag{26}$$

In formula (Eq. (24)), the ratio of the squares of the velocities can be represented as the identity

$$\frac{V^2}{V_\infty^2} = \frac{V^2}{a^2}\frac{a^2}{a_0^2}\frac{a_0^2}{a_\infty^2}\frac{a_\infty^2}{V_\infty^2} = \frac{M^2}{M_\infty^2}\frac{a^2}{a_0^2}\frac{a_0^2}{a_\infty^2}.$$

The use of isentropic formula (Eq. (13)) allows us to write the last equation in the form

$$\frac{V^2}{V_\infty^2} = \frac{M^2}{M_\infty^2} \cdot \frac{E(M_\infty)}{E(M)}.$$

Substitution of the found ratio of the squares of the velocities into the formula (Eq. (24)) for the pressure coefficient of the incompressible flow gives

$$c_p^0 = 1 - \eta_c^2 \frac{M^2}{M_\infty^2} \cdot \frac{E(M_\infty)}{E(M)}. \tag{27}$$

Using isentropic formulas for pressure ratios in the form

$$p_0/p = [E(M)]^{\frac{\kappa}{\kappa-1}}, \quad p_0/p_\infty = [E(M_\infty)]^{\frac{\kappa}{\kappa-1}},$$

we write down the pressure-drop coefficient of the compressible gas (Eq. (26)) as follows

$$c_p = \frac{2}{\kappa M_\infty^2}\left\{\left[\frac{E(M_\infty)}{E(M)}\right]^{\frac{\kappa}{\kappa-1}} - 1\right\}. \tag{28}$$

In order to calculate the pressure coefficient for a compressible gas from the known value of the incompressible flow coefficient, we introduce auxiliary functions proposed by G. Burago [20].

$$F(M) = [E(M)]^{\frac{\kappa}{\kappa-1}}, \quad G(M) = \frac{M^2}{E(M)} \cdot \left\{1 + [E(M)]^{\frac{1}{\kappa-1}}\right\}^{-2}. \tag{29}$$

Using the new functions, the formulas for the pressure-drop coefficients of an incompressible flow and a compressible gas can be represented in the form

$$G(M) - G(M_\infty)\left(1 - c_p^0\right) = 0, \tag{30}$$

$$c_p = \frac{2}{\kappa M_\infty^2}\left[\frac{F(M_\infty)}{F(M)} - 1\right]. \tag{31}$$

In order to recalculate the values of the pressure-drop coefficient of the incompressible flow to the pressure-drop coefficient of the compressible gas for a given value of the Mach number of the free-stream flow, it is necessary to solve the transcendental Eq. (30) with respect to the local Mach number, and then use the formulas (31) and (29) to calculate the pressure-drop coefficient of the compressible gas.

The dependence of the critical Mach number on the pressure-drop coefficient of the incompressible flow can be obtained. For this, it is necessary to solve Eq. (30) with respect to the condition $M = 1.0$, i.e. solve an equation of the form

$$G(1.0) - G(M_*)\left(1 - c_p^0\right) = 0. \tag{32}$$

3. Results and discussions

The presented method is approximate; therefore, it is necessary to demonstrate the consistency of this method with other calculation methods in order to analyze the accuracy of the calculations.

The calculation of the dependence of the critical Mach number on the pressure-drop coefficient of an incompressible flow using Eq. (30) is shown in **Figure 2**. **Figure 2** also shows the Khristianovich's curve [13].

Comparison of the two calculations shows good agreement. After the publication of the work of Burago [20] in 1949, a number of approximate methods based on the Chaplygin gas model were developed. Burago's method refers to the approximate mathematical models for accounting for the compressibility of the flow. The effectiveness of approximate mathematical models is always shown by comparison with experimental or calculated data for various bodies. Calculations carried out according to the method described above showed that this method is not inferior in accuracy to more rigorous methods, at the same time the advantage of this method over other methods of accounting for the flow compressibility in terms of speeds is undeniable.

Here are some comparisons. In **Figure 3** shows a comparison of the calculation by the method described above with the calculated data of L. Sedov [5] and by the Glauert formulas ([4], p. 311) and Kárman-Tsien ([4], p. 311). Glauert formula is

$$C_p = \frac{C_p^0}{\sqrt{1-M_\infty^2}}, \tag{33}$$

and the Kárman-Tsien formula is

$$C_p = \frac{C_p^0}{\sqrt{1-M_\infty^2} + 0,5\left[1-\sqrt{1-M_\infty^2}\right]C_p^0}. \tag{34}$$

In **Figure 3**, for the pressure coefficient, the experimental data are presented as points ([4], Figure 121) and a curve that is an approximation of these points ([5], p.388). Comparison analysis indicates that the calculation by the Glauert formula (34) is approximate and for the Mach number $M_\infty \geq 0,4$ it can be argued that the Glauert formula should not be used. The Kárman-Tsien formula (35) gives a better agreement with the experimental data than the Glauert formula, but for the considered pressure-drop coefficients ($C_p^0 = -0.6$ and $C_p^0 = -0.73$) it is inferior in accuracy to the calculated data of L. Sedov [5] and the data obtained by the author

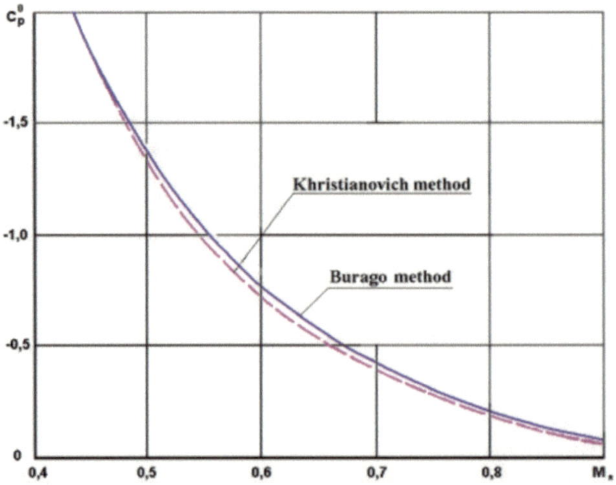

Figure 2.
Comparison of calculation results for critical Mach number based on Burago and Khristianovich methods.

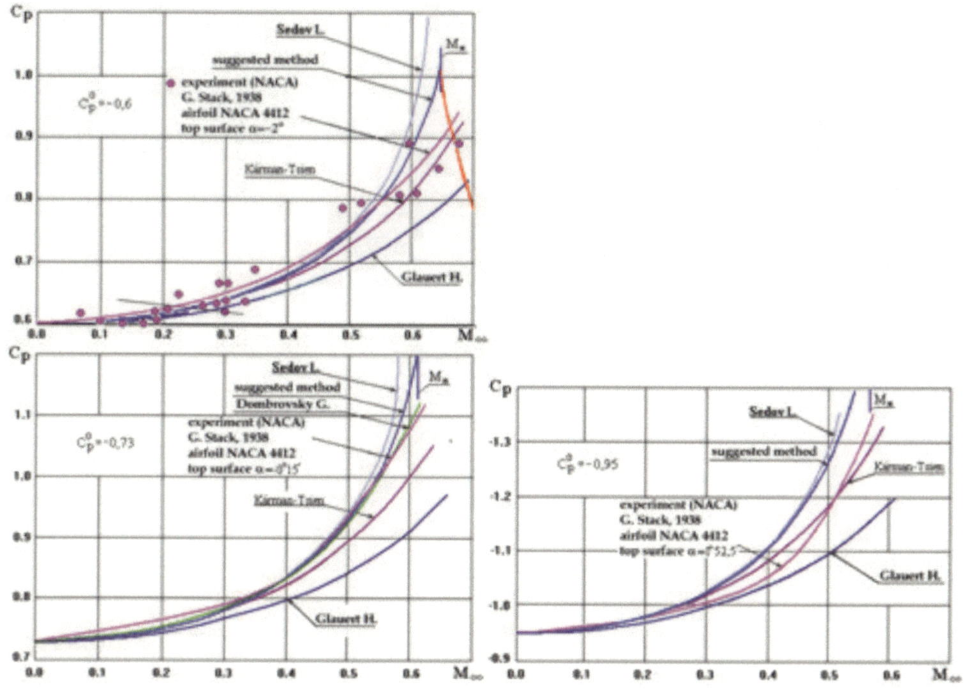

Figure 3.
Comparison of different methods account compressibility.

of the chapter by the method outlined above [1]. It should be noted that the calculated data by Burago's method for all three considered cases are closer to the experimental data of G. Stack [19] than the results of the calculation by the Sedov's method [5].

Accounting for gas compressibility according to the method described above and the Kárman-Tsien method [see. Formula (35)] is based on recalculating the pressure-drop coefficients of an incompressible liquid C_p^0 by the values of the pressure-drop coefficients of a compressible gas C_p for the given free-stream Mach numbers and ratio of specific heatsκ. An interesting question is: how much differ these two methods in a wide range of variation of Mach numbers and pressure coefficients for an incompressible fluid? **Table 1** compares these two methods based on the data in **Figure 3**, from which it follows that the maximum discrepancy increases with decreasing value C_p^0 and at $M_\infty \to M_*$. The maximum relative error of the two methods in the considered range of variation of Mach numbers and pressure coefficients reaches 15–16%.

The advantage of the Burago method over the Kárman-Tsien method is demonstrated in **Figure 4**, which presents the calculated data and experimental data of various authors for two bodies: an ellipse (**Figure 4a**) and a biconvex airfoil (**Figure 4b**) with relative thicknesses of 20% and 6% respectively. **Figure 4a** shows a comparison of the results of calculating the pressure drop coefficient on the surface of an ellipse of relative thickness δ = 0.2 in a compressible flow. The calculations are performed for the critical Mach number M_* = 0.7 (**Table 2**).

The calculated data using the Sells finite difference method (1968) are taken from [16]. It can be noted that the results of the Burago method and the Kárman-Tsien (34) are in good agreement to within approximate values $x \le 0.1$. For values of relative coordinate along the big axis of an ellipse $0.1 \le x \le 0.5$ the significant divergence between the results of the Kárman-Tsien formula (34) and Burago's

C_p^0	Mach number M_∞	C_p Kárman-Tsien	C_p Burago	δ %
−0.1	0.3	−0.1051	−0.1048	0.29
	0.6	−0.1266	−0.1269	0.24
	M_* = 0.886	−0.2289	−0.2165	6
−0.5	0.3	−0.5305	−0.5315	0.18
	0.6	−0.6667	−0.7042	5.0
	M_* = 0.679	−0.7489	−0.8696	14
−1.0	0.3	−1.0742	−1.0796	0.5
	0.4	−1.1432	−1.1659	2.0
	M_* = 0.558	−1.3427	−1.5874	15
−1.5	0.3	−1.6315	−1.6474	1.0
	0.4	−1.7566	−1.8320	4
	M_* = 0.486	−1.9245	−2.2790	16
−2.0	0.3	−2.2029	−2.2389	1.6
	0.4	−2.4009	−2.6067	8
	M_* = 0.437	−2.5034	−2.9683	16
−2.5	0.3	−2.7890	−2.8618	2.5
	M_* = 0.400	−3.0782	−3.6432	16
−3.0	0.3	−3.3904	−3.5221	3.7
	M_* = 0.371	−3.6515	−4.3127	15
−3.5	0.3	−4.0076	−4.2352	5
	M_* = 0.348	−4.2265	−4.9949	15
−4.0	0.3	−4.6414	−5.0245	8
	M_* = 0.329	−4.8020	−5.6830	15
−4.5	0.3	−5.2922	−5.9404	11
	M_* = 0.312	−5.3714	−6.3339	15

Table 1.
Comparison Burago's and Kárman-Tsien methods.

method is observed. For the values of the relative coordinate along the major axis of the ellipse $0.1 \le x \le 0.5$, there is a significant discrepancy between the Kárman-Tsien results and the Burago method. However, it can be noted that the calculations of other authors are in good agreement with the calculations by the Kárman-Tsien formula (34). The calculation according to this method approaches the value of the pressure coefficient, indicated in **Figure 4** as C_p^*, which corresponds to the critical Mach number. As can be seen from **Figure 4**, the values determined in the article [16] and by the author practically coincide. **Figure 4** also shows the distribution of the pressure-drop coefficient C_p^0 for an incompressible fluid. **Figure 4** shows the strong influence of the compressibility factor. Of greatest interest to researchers is the value of the critical Mach number for a cylinder in a transverse flow. In **Figure 5 (a, b)**, the results of calculations of the relative velocity on the surface of the cylinder, obtained using the proposed approximate technique, are compared with the calculations of other authors performed by other methods.

Integral relationship method [17], method of Legendre transformations [10], method approximation of adiabat [7], Rayleigh-Janzen method (Simisaki [17]; Shih-I

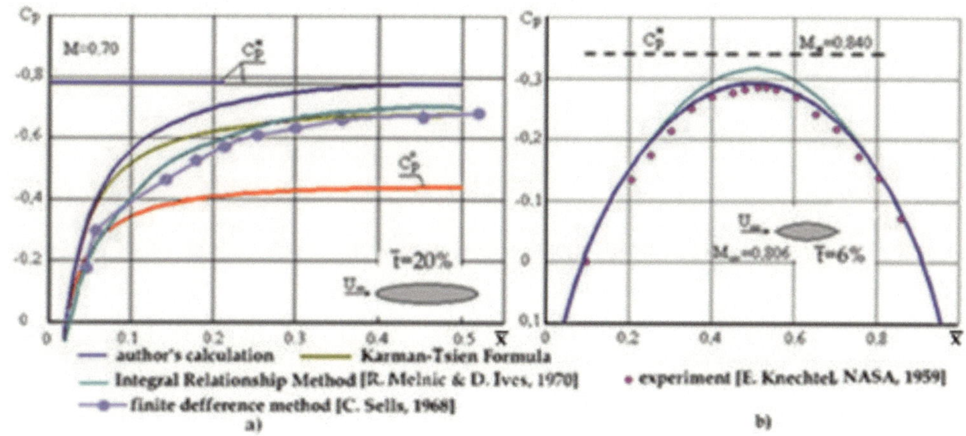

Figure 4.
Comparison of the calculated values of the pressure drop coefficient on the surface of the ellipse (relative thickness 20%) (a); for a biconvex airfoil (relative thickness 6%) (b).

M_*	Method and authors of calculation results	δ %
0.415	C. Jacob's calculation [7]	10
0.409	Poggi method. Calculations by S. Kaplan [12]	9
0.404	I. Imei's calculation	8
0.400	Rayleigh-Janzen method. T. Simisaki's calculations	7
0.399	The method of integral relations by A.A. Dorodnitsyn. calculations by P.I. Chushkin	7
0.39853¤0.00002	The multi-layer method of integrated relations by A.A. Dorodnitsyn. calculations by R. Melnik and D. C. Ives	7
0.3983¤0.0002	The Rayleigh-Janzen method. Hoffman's calculations	7
0.396	The method of approximation of adiabat. Approximation A4. G. A. Dombrovsky	6
0.390	The method of approximation of adiabat. A3 approximation. G. A. Dombrovsky	5
0.390	The method of integral relations by A.A. Dorodnitsyn. Calculations by M. Holt and B. Messon	5
0.390	Calculations by I. Imei. Z. Hashimoto	5
0.37170¤0.00001	author's calculation	—

Table 2.
Comparison of the calculated values for the critical Mach number on the surface of the cylinder.

Pai [6]) are compared in **Figure 5**. The results of Simisaki's calculations were taken from [17]. The calculations were carried out for the Mach number almost equal to the critical Mach number $M_\infty \approx 0.372$ (**Table 2**). It can be noted that in **Figure 5**, the author's calculation is in good agreement with the data of other authors for velocities on the cylinder surface, taking into account the compressibility factor.

Let us consider the numerical values of the critical Mach numbers M_* for a flow around a circular cylinder obtained by different authors. **Table 1** shows a comparison of the results of calculating the critical Mach numbers by different methods. The value δ denotes the relative error of comparing the values of the critical Mach numbers calculated by different methods and by this method.

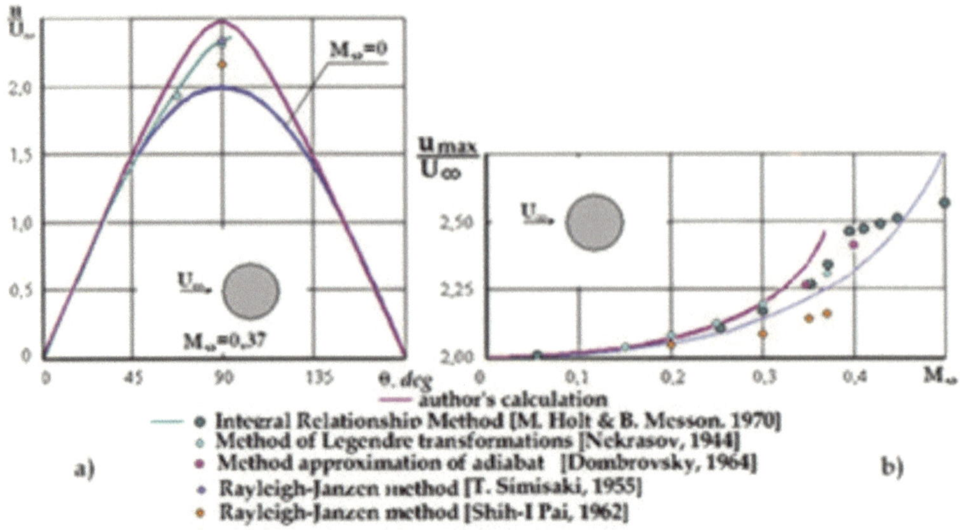

Figure 5.
The velocity on the cylinder surface vs. Mach number.

If we exclude from the comparison the early work of Kaplan [12] in which the adiabatic exponent for air was taken equal $\kappa = 1.408$ which affected the value of the calculated critical Mach number and the work of Jacob (see [7]) then based on the data in **Table 1** it can be argued that the accuracy of calculating the critical Mach number on the surface of the circular cylinder by the Burago's method in comparison with other methods on average approximately corresponds to a relative error of 5%. From **Table 1** it can be seen that the critical Mach number on the surface of the circular cylinder calculated by Burago's method is the smallest of the data presented in the **Table 1**. This does not mean that this value is the most inaccurate compared

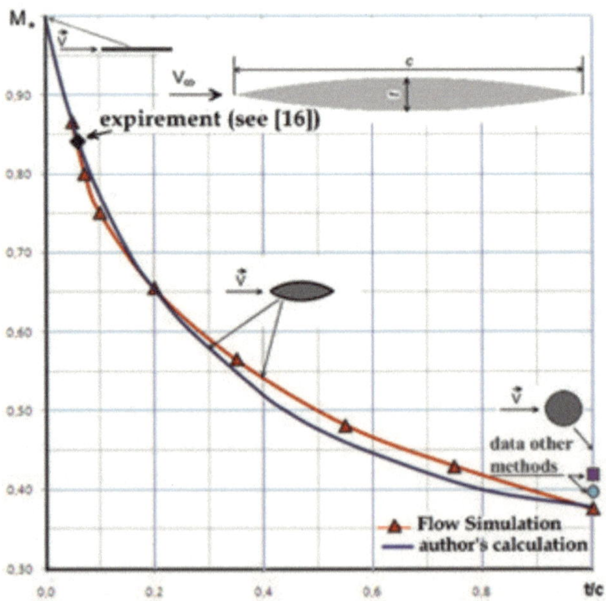

Figure 6.
Critical Mach number vs. relative thickness of the biconvex airfoil.

to other methods. Holt and Messon [17] point out that "symmetric flows (in which a shock does not occur) were calculated only up to the Mach number of the free-stream flow M_{∞}=0.37". This circumstance indirectly confirms that the exact value of the critical Mach number is closer to that determined by the Burago method than by other methods.

An important question is the question: how does the viscosity of the medium affect the value of the critical Mach number? **Figure 6** shows a comparison of the calculated data for the flow around the biconvex airfoil using the ideal gas and viscous models. **Figure 6** also shows the experimental results that can be observed also in **Figure 4(b)**.

Calculations for viscous gas are performed in a computational package Flow Simulation (SolidWorks). The calculations in the Flow Simulation software are based on solving the Navier–Stokes equations. The calculation domain and the grid used in the calculations Flow Simulation are shown in **Figure 7(a)** and **(b)**. The flow velocity was calculated at a distance equal to the thickness of the boundary

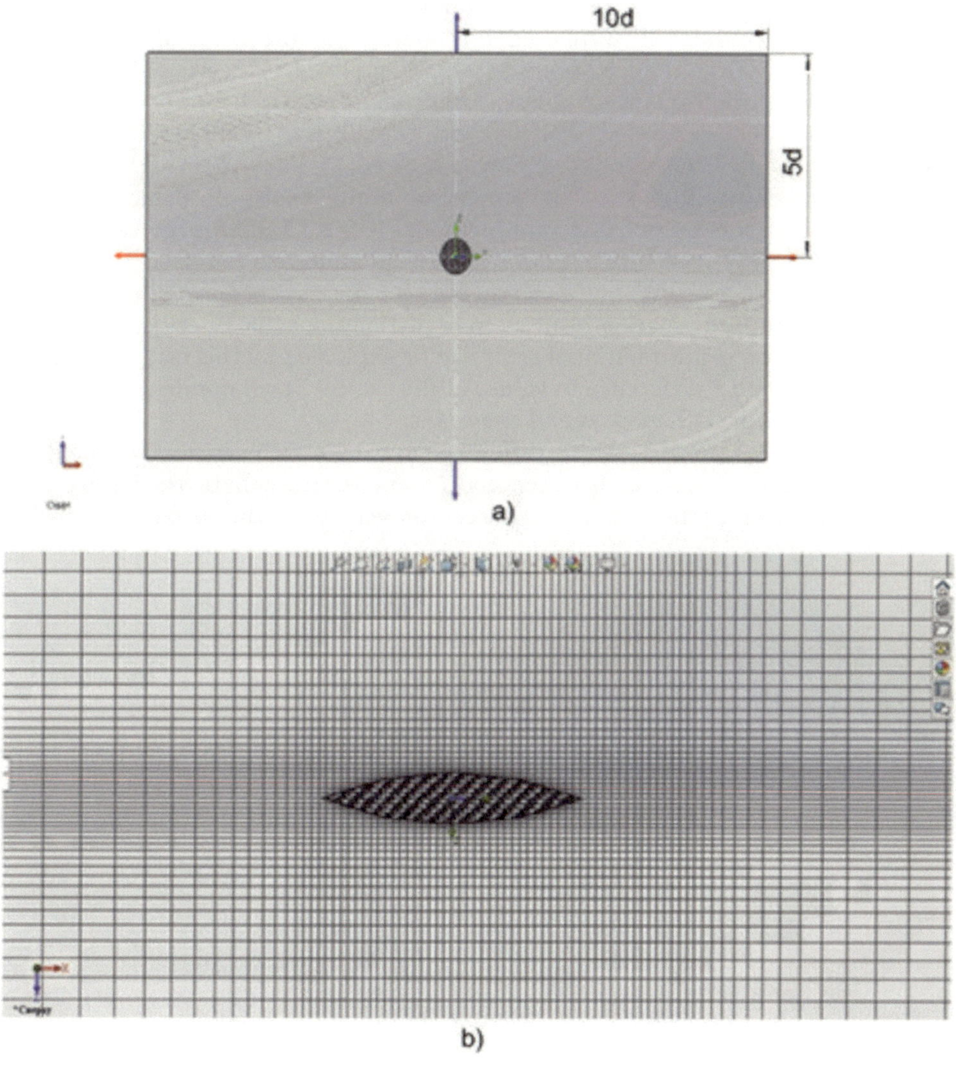

Figure 7.
The calculation domain and grid used in package flow simulation.

layer. In the calculations, the free-stream velocity varied, and as soon as the local flow velocity at the boundary of the layer boundary became equal to the sound velocity, the critical Mach number was calculated.

Based on **Figure 6**, it can be concluded that viscosity has practically no effect on the value of the critical Mach number.

Another two-dimensional body, the flow around which is well studied is an ellipse. **Figure 8** shows the calculated results for the critical Mach number depending on the degree of compression (relative thickness, $p = a/b$) of the ellipse. **Figure 8** also shows the value of the critical Mach number for a two-dimensional body formed by two contacting cylinders. Calculations of the critical Mach number for such a 2-D body were first performed by the author [1].

We have looked at comparing calculations for two-dimensional bodies. The high accuracy of the calculation of the critical Mach numbers by the Burago method is shown.

Let us now consider the applicability of the Burago method for axisymmetric flows. It is also of interest to compare the critical Mach numbers for two-dimensional and axisymmetric flows. Ellipses and ellipsoids of revolution (spheroids) with various factor of compression δ are chosen. For calculations the well-known potential models for ellipses and spheroids [21] were used. It is obvious that the maximal error will correspond to flow around thick bodies ($\delta \to 1.0$) and for the case $M_\infty = M_*$. In **Table 3** and in **Figure 9(a)** results of calculation of critical Mach number for ellipses and spheroids with an offered method and with Dorodnitsyn method of integral relations executed by Chushkin [14, 15] are compared.

Table 3 and **Figure 9(a)** shows a comparison of the results of calculating the critical Mach numbers by two different methods. **Table 3** and **Figure 9(a)** shows a good agreement. It is necessary to notice that calculations in a range of $(0.4 \le \delta \le 0.8)$ were executed by Chushkin only for the second approximation and each subsequent approximation resulted in reduction the value of critical Mach number. Therefore, some additional error can be explained by this fact. It can be noted that the relative difference in values of the critical Mach numbers increases with increasing δ for both ellipses and spheroids.

However, the maximum relative difference in values of the critical Mach numbers for the elliptic cylinder does not exceed 7% and for the sphere 8%. **Figure 9(b)** shows a comparison of the ratio of the maximum velocity at the surface of an

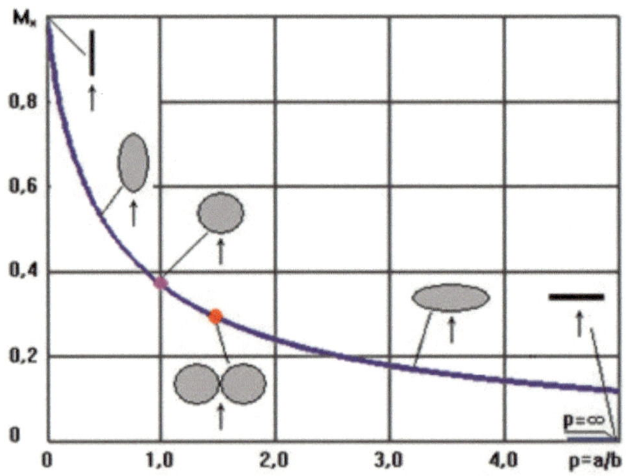

Figure 8.
Critical Mach number vs. relative thickness of the ellipse.

δ	Ellipses				Ellipsoids			
	$C^0_{p\ min}$	Chushkin [14]M_*	Calculation M_*	Relative difference %	$C^0_{p\ min}$	Chushkin [14] M_*	Calculation M_*	Relative difference %
0.05	−0.103	0.869	0.884	1.8	−0.014	0.984	0.980	0.4
0.10	−0.210	0.803	0.807	0.5	−0.042	0.957	0.945	1.3
0.15	−0.323	0.752	0.748	0.5	−0.080	0.929	0.905	2.6
0.20	−0.440	0.709	0.700	1.3	−0.122	0.899	0.868	4.0
0.40	−0.960	0.588	0.566	3.7	−0.337	0.783	0.742	5
0.60	−1.560	0.506	0.480	5	−0.602	0.692	0.648	6
0.80	−2.240	0.447	0.418	6	−0.908	0.620	0.576	7
1.00	−3.000	0.399	0.372	7	−1.250	0.563	0.519	8

Table 3.
Critical Mach number at a flow around ellipses and spheroids.

elliptical contour to the velocity at infinity as a function of the Mach number (symmetric flow). **Figure 9(b)** uses the relative ellipse thickness δ = 0.1. Here, along with the calculations by the proposed method, the results of calculations obtained by the theory of small perturbations and by the Hantz and Wendt second approximation (see the Book [7]), as well as the results of calculations obtained by the Dombrovsky method [7] are shown. It can be noted that the results of calculations using the proposed method practically coincided with the results of Dombrovsky. The critical Mach number equal to $M_* \approx 0.807$ (**Figure 9(b)**) was obtained by the Dombrovsky method of approximation of the adiabat. This is in very good agreement with Lighthill's data (see Book [7]) and Dombrovsky's result [7] $M_* \approx 0.81$ and Chushkin's result $M_* \approx 0.803$ (**Table 2**). Kaplan's result [12] $M_* \approx 0.857$ should be considered less accurate, but the relative error does not exceed 7%.

Also, it follows from **Figure 9(a)** that critical Mach number for the two-dimensional case is always less than for axisymmetric case for bodies with the same cross-section.

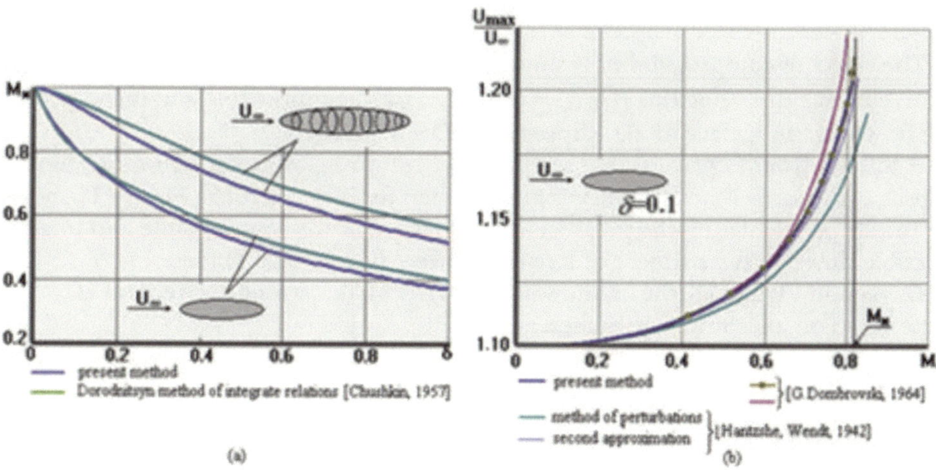

Figure 9.
(a) Critical Mach number vs. relative thickness of an ellipse and ellipsoid; (b) relative maximal velocity of the flow around ellipse vs. Mach number.

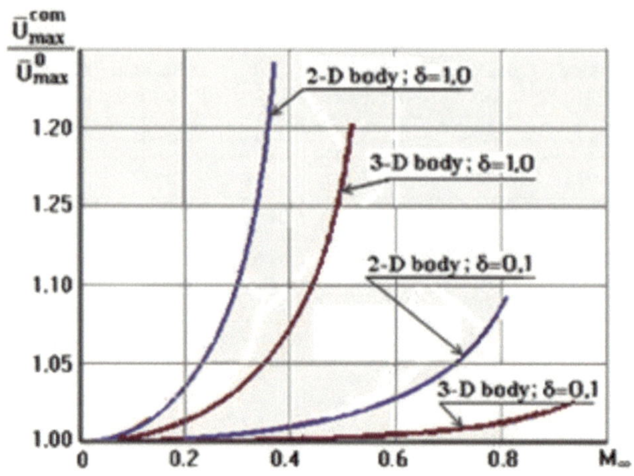

Figure 10.
Maximal velocities on surface of the ellipse and spheroid vs. Mach number.

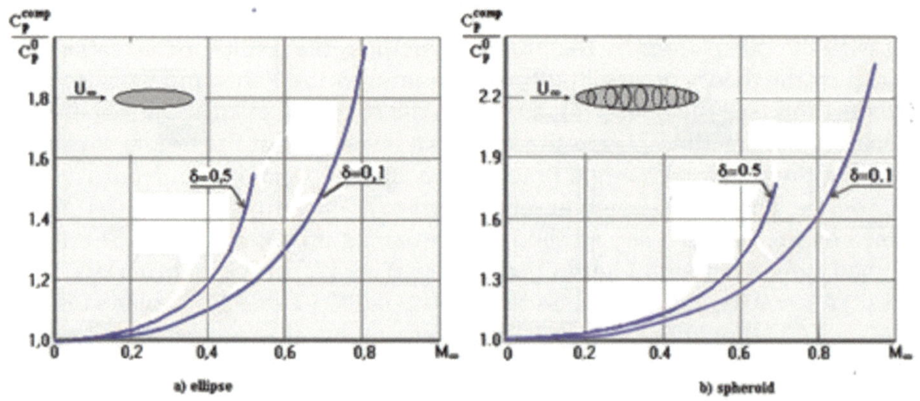

Figure 11.
Influence of Mach number of compressible flow on pressure-drop coefficient on the surface of the an ellipse and spheroid.

The effect of compressibility is shown in **Figure 10**, which shows the ratio of the relative maximum velocities $\left(\overline{U}_{\max} = U_{\max}/U_{\infty}\right)$ for compressible and incompressible flows on the surface of 2D ellipses and 3D spheroids.

It follows from **Figure 10** that the effect of compressibility for two-dimensional ellipses at the same Mach number is greater than for 3D-spheroids. **Figure 11** shows the results of calculating the pressure-drop coefficient for compressible and incompressible flows on the surfaces of two- and three-dimensional bodies.

As seen in **Figure 11**, the compressibility effect has a stronger effect on 2D bodies than on axisymmetric bodies.

4. Conclusions

This paper presents the results of calculating the critical Mach numbers of the flow around two-dimensional and axisymmetric bodies. A sufficiently high accuracy of calculating the critical Mach numbers for engineering calculations using the proposed method is shown. The method allows one to determine the parameters of a compressible flow from the values of the flow of an incompressible fluid up to a speed corresponding to the critical Mach number. This method does not depend on the means determination parameters of the incompressible flow. The calculation in software Flow Simulation was shown that the viscosity factor does not affect the value critical Mach number. It was found that with an increase in the relative thickness of the body, the value of the critical Mach number decreases. It was also found that the value of the critical Mach number for the two-dimensional case is always less than for the axisymmetric case for bodies with the same cross-section.

Conflict of interest

The author declares no conflict of interest.

Acronyms and abbreviations

AMM	Applied Mechanics and Materials
NACA	National Advisory Committee for Aeronautics
TsAGI	Central Aerohydrodynamic Institute
TM	Technical Memorandum
TR	Technical Report
HSH	High Speed Hydrodynamics

Appendices and nomenclature

u, v	velocity components along axis x и r;
ρ	local density of gas;
τ, σ	special functions (Eq. (2));
p	the static pressure;
∞	subscript indicates flow parameters at infinity;
0	lower index, specifies parameters of the flow in the stagnation point or upper index and specifies parameters of the flow of the incompressible liquid;
κ	the ratio of specific heats;
$P(p)$	the pressure function;
h	the enthalpy;
$\mathbf{V}$	the vector of full local velocity of flow;
M_∞	the Mach number at infinity;
M	the local Mach number;
$\overline{u} = u/U_\infty, \ \overline{v} = v/U_\infty$	dimensionless velocities;
$E(M)$	the function (Eq. (11));
a	the sound speed;
η_c	the compressibility factor of the flow (Eq. (15));
ε	given accuracy of calculation;
$\overline{U}$	local total relative velocity (Eq. (23));
c_p^0	the pressure-drop coefficient for an incompressible flow (Eq. (28));
$G(M), F(M)$	the functions (Eq. (30)).

Author details

Vladimir Frolov
Samara National Research University, Samara, Russia

*Address all correspondence to: frolov_va_ssau@mail.ru

References

[1] Frolov V. High-speed flows of the compressible fluid around two circle contours with a pair of symmetric vortices. In: Proceedings of the International Summer Scientific Scholl «High Speed Hydrodynamics» (HSH-2002); 16–23 June 2002; Cheboksary. Russia/Washington. USA: Computational Publications. 2002. p. 331–338

[2] Frolov V. High-speed compressible flows about axisymmetric bodies. In: Proceedings of the Fifth International Symposium on Cavitation (Cav2003). 1–4 November 2003. Osaka. Japan: CD-ROM. Cav03-OS-7-004. p.1–9

[3] Frolov V. Critical Mach number for flow around circular-arc airfoil. AMM; 2015: 798: 513–517. doi:10.4028/www.scientific.net/AMM.798.513

[4] Liepmann H., Puckett A. Introduction to aerodynamics of a compressible fluid. 1st ed. NY: Wiley; 1949. 331 p.

[5] Sedov L. Two-dimensional problems of hydrodynamics and aerodynamics. 3rd ed. Moscow: Nauka; 1980. 448 p.

[6] Shih-I Pai. Introduction to the theory of compressible flow. Literary Licensing. LLC. 2013. 400 p.

[7] Dombrovsky G. Method approximation of adiabat in theory 2-D flows of the gas. Moscow: Nauka. 1964. 158 p. (in Russian).

[8] Chaplygin S. Gas jets. Moscow: publishing house of Moscow University. 1902. 121 p. [Gas jets. Washington. 1944. NACA TM. 1063. 112 p.]

[9] Khristianovich S. Flow of bodies by gas at high subsonic speeds. Moscow: Publications TsAGI. 1940. No. 481. (in Russian)

[10] Nekrasov A. Two-dimensional gas motion with subsonic velocities. J. Applied Mathematics and Mechanics. 1944. Vol. VIII. No.4. p.249–266 (in Russian)

[11] Kudriashov L. Plane parallel gas flow past an ellipse. J. Applied Mathematics and Mechanics. 1947. Vol. XI. No.2. p. 119–128 (in Russian)

[12] Kaplan C. Two-dimensional subsonic compressible flow past elliptic cylinders. NACA Report. 1938. 624. p.245–252

[13] Khristianovich S., Yuriev I. Subsonic gas flow past a wing profile. Applied Mathematics and Mechanics. 1947. Vol. XI. No. 1. p. 105–118 [Subsonic gas flow past a wing profile. Washington. 1950. NACA TM. No.1250. 112 p.]

[14] Chushkin P. Flow of ellipses and ellipsoids by subsonic flow of gas. Computing Mathematics. collection 2. 1957. Moscow: Publishing House of Academy of Sciences USSR. p. 20–44 (in Russian)

[15] Chushkin P. Calculation of flow around arbitrary airfoil and bodies of rotation in subsonic flow of gas. Computing Mathematics. collection 3. 1958. Moscow: Publishing House of Academy of Sciences USSR. p. 99–110 (in Russian)

[16] Melnic R., Ives D. Calculation subsonic flows around flat airfoils with helps multi-layer method of integrated relations. In: Numerical Methods in Mechanics of Liquids. Moscow: Mir; 1973. p. 26-38 (translated in English)

[17] Holt M., Messon B. Calculation of flow of bodies with high subsonic speed with help method of integrated relations. In: Numerical Methods in Mechanics of Liquids. Moscow: Mir; 1973. p. 39-48 (translated in English)

[18] Hida K. On Subsonic flow of compressible fluid past prolate spheroid. Journal of the Phys. Soc. of Japan. 1953. Vol. VIII. No. 2, p.257 -264, arXiv: https://doi.org/10.1143/JPSJ.8.257

[19] Stack J., Lindsey W., Littell R. The compressibility burble and of compressibility on pressure and forces acting on an airfoil. NACA TR. 1938. 646. p.73-96

[20] Burago G. Theory of airfoils by accounting for compressibility of air. Moscow: Publishing house "Military-Air Engineering Academy of name of the prof. N. E. Zhukovsky". 1949. 166 p. (in Russian)

[21] Roze N., Kibel I., Kochin N. Theoretical hydrodynamics: Part 2. Leningrad. Moscow: The Main Edition of the Engineering-Theoretical Literature. 1937. 507 p. (in Russian)

Section 5

Manufacture

Chapter 10

Laser Based Additive Manufacturing Technology for Fabrication of Titanium Aluminide-Based Composites in Aerospace Component Applications

Sadiq Abiola Raji, Abimbola Patricia Idowu Popoola, Sisa Leslie Pityana, Olawale Muhmmed Popoola, Fatai Olufemi Aramide, Monnamme Tlotleng and Nana Kwamina Kum Arthur

Abstract

Titanium aluminides has the potential of replacing nickel-based superalloys in the aerospace industries because its density is almost half that of nickel-based alloys. Nevertheless, the room temperature properties (ductility) have made the wider application of this class of intermetallic alloy far from being realized. This has led to various research been carried out in adjusting the production processing and/or material through alloying, heat treatment, ingot metallurgy, powder metallurgy and most recently additive manufacturing processing. One of the additive manufacturing processing of titanium aluminide is laser engineered net shaping (LENS). It is used to produce components from powders by melting and forming on a substrate based on a computer-aided design (CAD) to shape the components. This contribution will focus on the laser processing of titanium aluminides components for aerospace applications. Also, the challenges confronting this processing techniques as well as suggested finding to solve the problems would be outlined. The objective of this work is to present an insight into how titanium aluminides components have been developed by researchers with emphasis on aerospace applications.

Keywords: additive manufacturing, titanium aluminide, composites, laser additive manufacturing (LAM), aerospace components

1. Introduction

Aerospace engine components are exposed to high temperatures as a result of enormous heat generation in the engine and formation of large temperature rise across the structural parts. These components are generally expected to possess a high strength-to-weight ratio in order to give high thrust, improved fuel efficiency at minimal fuel consumption while increasing speed of flight. Aircraft engines are usually fabricated by titanium alloys, nickel-based super alloys, superior steels or ultrahigh-temperature ceramics. Moreover, an important engineering material for manufacturing of gas turbine and aircraft engine components are alloys of titanium (Ti) because of good properties like corrosion resistance, excellent specific strength and most especially their low density [1]. Also, due to the biocompatibility of these alloys, they have been applied in biomedical industries as implants [2–4] and prosthetic beak [5]. The poor wear performance and low hardness of titanium alloys constrained their adoption in fabricating components to be subjected in contact forces of friction and wear [3, 6]. Likewise, titanium alloys have not been impressive for load-bearing components (LBC) in the aerospace industry because of uneven stress distribution on parts [7].

Beta titanium (β-Ti) alloys is reported to have been used to fabricate landing gears in Boeing aircraft from the Ti-5Al-5Mo-5V-3Cr because of better strength-toughness mishmash in β-Ti alloys compared to α + β and α alloys [8]. Titanium intermetallic alloys have great potentials in aerospace, automotive and power plants especially titanium aluminides (TiAl) because of their excellent creep and light-weight in relation to nickel-based superalloys [9]. The most feasible TiAl alloys with engineering interest are those of 44–48 at.% Al content because of the eutectoid transformation reaction occurring during solidification and cooling process [10]. **Figure 1** displays the phase diagram of Ti-Al between 0 and 60 at. % aluminum. Titanium matrix composites (TMCs) reinforced with ceramics has increased due to improvement in heavy load-bearing, wear and friction performance in service due to material efficiency with lower cost of parts fabrication.

A possible replacement of Ni-based superalloys is gamma-titanium aluminide (γ-TiAl) considered in manufacturing aero-engine and automobile engine parts because of their high strength, high stiffness, light weight good oxidation and corrosion resistance [11–20]. The choice of γ-TiAl is hinged on the fact that its density is almost half of Ni-based superalloy and high temperature creep properties [21, 22] with application temperature up to 750°C [23]. These has made γ-TiAl to be favorably accepted in aerospace industry. However, aero engine component like turbocharger operate at temperatures between 700 and 950°C which presents a concern of oxidation reaction apart from its room temperature ductility [22]. γ-TiAl

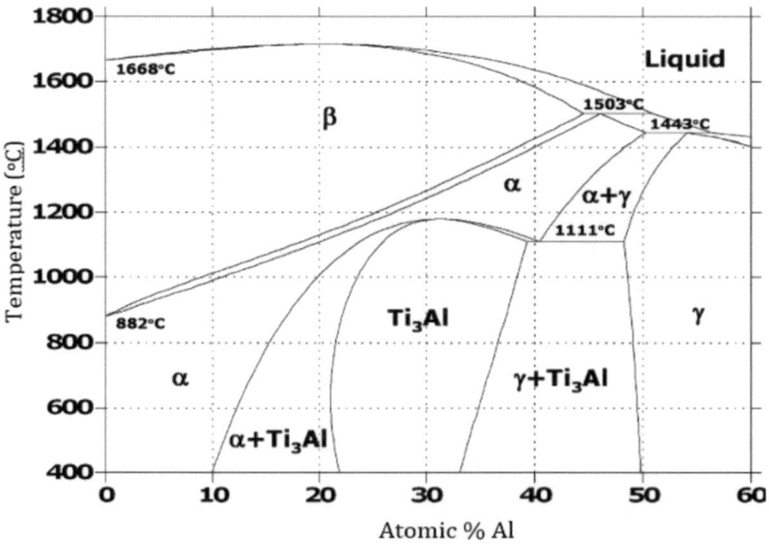

Figure 1.
Ti-Al binary phase diagram.

of 48 at.% Al generally comprise of a γ-phase and α_2-phase with $L1_o$ and DO_{19} structure respectively (major: γ-TiAl, minor: α_2-Ti_3Al) typically referred to as γ-TiAl [10, 24]. This γ-TiAl-based alloy are continuously been improved upon in the manufacturing nozzles, low pressure turbine blade (LPTB) and turbochargers in aerospace engines [25].

General Electric has applied the Ti48Al2Cr2Nb alloy called the GE4822 to manufacture the Boeing-787 Dreamliner's engine known as GEnx-1B that reduces both fuel consumption and CO_2 emission by 15% [11, 21, 26]. Due to the GE4822 alloys high-temperature capability, there has been works on improving the ductile property of γ-TiAl which have been used for low-pressure turbine (LPT) blades [11, 16, 18, 27] and to expand the temperature range 800–850°C [11]. High Nb-containing TiAl known as TNM have been produced with higher yield strength, oxidation resistance and better creep than conventional TiAls [15, 19, 23, 26, 28, 29].

However, the extremely poor tensile ductility (<1% elongation) of TiAl-based alloys at room temperature, poor fracture toughness and high-temperature poor workability makes it difficult to fabricate and strongly limits their applications in the industry [10, 13, 15, 16, 18, 30]. This has made conventional processing techniques very problematic in manufacturing TiAl-based components. Hence, a lot of interest in improving the ambient temperature characteristics and oxidation resistance at elevated temperature, methods such as alloying and post-processing heat treatment has been suggested to help advance its properties.

Apart from the potentially improved geometrical complexity, additive manufacturing (AM) permits manufacturing of innovative materials with functionally graded capabilities during fabrication [14]. The applicability of TiAl alloys is becoming more progressive due to the growing knowledge of the connections concerning mechanical properties and microstructural evolution as well as modifications in processing operations [18]. However, even with AM technologies issues of the mass production quantities is still not making the technique cost-effective [31].

Currently, AM has gained acceptance for manufacturing customized products with complex geometrical freedom, more homogeneous microstructure and functionally enhanced parts gradually replacing traditional processes [14, 28, 32]. Due to this property, it is possible to manufacture components and finished parts which could not be implemented with conventional manufacturing technologies or only at great expense. AM routes is poised to produced γ-TiAl parts for aerospace applications.

Laser additive manufacturing (LAM) is a class of AM process that makes use of high-powered laser beam in melting and fabricating metal powders into three-dimensional (3D) components from a predesigned computer-aided design (CAD) model. One of the highly desirable LAM techniques is laser engineered net shaping (LENS®) regarded for producing ceramics, composites and functionally graded materials (FGMs) which are difficult to produce [33, 34].

The objective of this work is to give an overview of LAM of TiAl alloys and composites while presenting a review of crucial matters, investigation tendencies, fabrication and progress achieve on γ-TiAl based materials for aerospace. A succinct overview of TiAl was first presented followed by AM of titanium alloys particularly those applied in aerospace industries. The paper subsequently highlights a comprehensive appraisal of latest studies on γ-TiAl laser processing and post-processing effects on microstructural evolution and mechanical properties. Prospects, challenges and suggested findings were discussed at the end of this work.

2. Additive manufacturing (AM)

AM is a material joining process from a three-dimensional (3D) CAD model data to build an object, in typically layer by layer [32, 35]. AM has been known by names which includes freeform fabrication (FFF), additive layer manufacturing (ALM), rapid prototyping (RP) and 3D-printing (3DP). AM have been widely adopted in industries especially the aerospace and medical field with recent application in radio-frequency (RF) for microwave and millimeter wave devices [32]. The distinguishing feature of this technology is ability to manufacture objects through layer upon layer building [36] and consistent component functions requiring little or no additional process [37].

The two key classes of AM technologies are direct energy deposition (DED) and powder bed fusion (PBF) as depicted in **Table 1**. The DED methods offers prospects in engineering applications for producing functional parts with intricate geometries from metallic powders [31]. The powder particles used are preferably spherical with the DED being more sensitive in size distribution and wires are adopted as precursor for certain DED processes for higher production rate [38]. The process melts the material feedstock (powder or wire) using a high-powered laser beam. The most commonly used lasers are CO_2 and ND: YAG which has been used for fabricating complex and customized parts, prototyping of metallic parts and repairing/cladding of components [31]. This method is ideal to directly make components with limited tooling but the produced parts exhibits relatively low fracture toughness and prone to defects like cracks and pores due to fast heating and cooling rates involved [39, 40].

In the PBF techniques, a heating source (laser or electron beam) scans metallic powders on a preplaced build platform to melt the powders in a predefined path. This is controlled based on CAD model to build the object in a layer by layer tool path. It has the benefits of reduced lead time and material waste [41] and control of melt pool dynamics and stability [42].

There have been remarkable improvements in AM primarily by the Ti alloys production for aerospace components. Though, PBF is competent to make near-net shaped objects and DED also can be applied to refurbish/repair parts and modify features apart from fabricating 3D objects. The high-temperature gradient due to focused instant heating and cooling is the main difficulty encountered by these techniques.

The majority of investigations on AM materials demonstrates that their mechanical properties of better than the traditionally manufactured titanium alloys. The principal application of additive manufactured Ti-alloy is the aerospace

Classes	Technology	Source of energy	Material used
Powder bed fusion (PBF)	Electron beam melting (EBM)	Electron beam, laser beam	Metal, ceramics and polymers
	Selective laser melting (SLM)		
	Direct metal laser sintering (DMLS)		
	Selective laser sintering (SLS)		
Directed energy deposition (DED)	Laser engineered net shaping (LENS)	Laser beam	Metals (powder and wire)
	Laser cladding		
	Electronic beam welding (EBW)		

Table 1.
The major classes of additive manufacturing (AM).

industry followed by the medical industry while its gradually been accepted in automotive industry. AM presents encouraging platform to produce highly effective Ti-alloy component with intricate geometry [43, 44]. Whereas, these alloys are continuously been studied, the problem encountered is due to their additional processing requirements.

Laser-based AM, however, achieves the built parts defined in a CAD model by melting and solidifying powder on a layer by layer basis [45]. This method is known to have been adopted in aerospace to manufacture and create components of graded material using Ti-based alloys. The LAM technologies produce fully dense parts from wire or powder feedstock [46, 47]. It has the merit of cost and time saving in comparison to traditional methods of casting and forging. A pertinent problem encountered in LAM when processing titanium matrix composites form various literatures is the unevenness of particle distribution. Also, the appearance of non-melted particles leading to reduced plasticity.

2.1 Additive manufacturing of titanium alloys

One of the titanium alloys that are ideal for aerospace applications is the β-Ti alloy because it has a combined property of high strength and lightweight [48]. Titanium alloys have proven to be extremely attractive and important materials in aerospace applications because of excellent combinations of outstanding corrosion and mechanical properties [39, 47, 49, 50]. Recently, there has been increased interests in the development of additively manufactured titanium (and their alloys) parts in industries. However, the widespread applications are mitigated by expensive processing involved compared to some material of similar function [39].

A lot of studies have shown that Ti-alloys produced by AM possess considerable better mechanical properties than the conventional materials. Also, recent development has led to increased acceptance of the technologies, contributing new potential concerning functional incorporation and lightweight components.

Saboori et al. [31] presented a review of AM on titanium components by direct laser deposition (DLD) analyzing the significance prospects of DED process. The connection among titanium alloys microstructural characteristics and process parameters were outlined. It was understood that DED processed parts undergo complex thermal history severely induced due to process uncertainties and process parameters. Even with attempts in optimizing process parameters to curtail flaws, enhance mechanical properties and microstructure of the component, there still existed the challenge of striking a balance. However, titanium parts fabricated by DED still had increased strength and poor ductile properties. It was stated that this alloy suffer anisotropy in the tensile characteristics because components fabricated horizontally demonstrate increased ductile strength compared with those build vertically.

In the work of Hu et al. [3], titanium matrix composites (TMC) has been fabricated by LENS using a 3D quasi continuous network (3DQCN) and TiB as reinforcement. The author examined the influence of energy input and reinforcement on parts quality and wear resistance. Superior wear performance was recorded as a result of TiB addition as a reinforcement. In a related work by Hu et al. [6], the author examined the influence of laser power and TiB reinforcement on mechanical properties. It was shown that laser power greatly affects the resulting microstructures but the strengthening of TiB increases the ultimate compressive strength (UCS) and microhardness of TiB-TMCs. **Table 2** shows influence of laser power in varying microhardness of LENS processed TiB-TMCs.

Also, the mechanical properties and defects (porosity) of TiB-TMCs parts fabricated using an innovative ultrasonic vibration-assisted (UV-A) LENS have been

Materials	Laser power	Microhardness ($HV_{1.0}$)	Processing
Cp-Ti	200 W	345.5 ± 5.4 [3]	LENS
Cp-Ti	125 W	304 ± 10 [6]	LENS
TiB-TMCs	200 W	392.6 ± 8.9 [3]	LENS
TiB-TMCs	125 W	339 ± 9 [6]	LENS
TiB-TMCs	200 W	393 ± 9 [6]	LENS
TiB-TMCs	200 W	405–429 [52]	UV-A LENS
TiB-TMCs	300 W	428–488 [52]	UV-A LENS

Table 2.
Microhardness comparison of TiB-TMCs.

investigated by Ning, Hu and Cong [51]. The procedure adopted exhibited remarkable influences on porosity, pore size, TiB whisker size and distribution, QCN microstructural grain size and microhardness. Consequently, fine microstructures were obtained due to refinements of both TiB whiskers and QCN microstructural grains UV-A LENS providing larger area of grain boundaries to hinder the dislocation motion, thus improving the microhardness of fabricated parts.

A research was conducted by Grove et al. [50] to investigate and compare the surface integrity and machinability of Ti-5553 alloy with composition Ti-5Al-5V-5Mo-3Cr produced by casting, SLM and SLM + heat treatment (in-situ). Higher tool wear was noticed in the in-situ heat treated alloy while the tool wear for the other two does not seem to differ. Therefore, it proves titanium alloys' machinability is extremely affected further processing methods.

According to Uhlmann et al. [52], the aviation industry requires a drastic reduction in NOx and CO_2 emission in the coming years ahead. The authors' study was focused on highly effective and efficient titanium alloys with optimizing process parameters for structures manufactured by the SLM machine. The post-processing of TiAl6V4 parts fabricated by SLM display vast possibility of enhancing surface quality. Examination of computed tomography and microstructure confirms that porosity of titanium alloy parts can be minimized through HIP-process.

Silica ($SiCO_2$) coatings have been created on commercially-pure titanium (Cp-Ti) by Heer and Bandyopadhyay [53] using LENS with subsequent stress relieve heat treatments and post-deposition laser passes. Ti_5Si_3 phase was formed leading to high hardness of ~1500 HV in the coatings and wear rate was mostly reduced in comparison to Cp-Ti irrespective of heat treatment as well as rise in laser pass commonly reduce wear rates. The microstructures revealed dendrite shaped and morphology of columnar deposit around primarily matrix of α-titanium.

A process of reactive deposition AM was adopted by Traxel and Bandyopadhyay [45] for maintaining Ti-Zr-BN composite parts using commercially pure titanium (Cp-Ti) with addition of Zr and BN. This was aimed at improving elevated temperature and wear resistance abilities if TMC using LENS. The as-fabricated BN-containing components shows TiB_2, TiN and TiB as reinforcement with SEM image of CpTi-BN and CpTi-Zr-BN presented in **Figure 2** below. Zr-addition displays a combination of composites of composites and alloys leading to high hardness and ultimate compressive strength (UCS) with enhanced wear resistance. This method is expected to be used in creating novel coatings and structures from vast powder feedstocks to develop the bulk and surface characteristics of titanium related metals.

Effects of thermal history and build direction were studied by Mantri and Banerjee [48] on β-Ti alloys containing Mo-12 wt% and V-20 wt%, produced

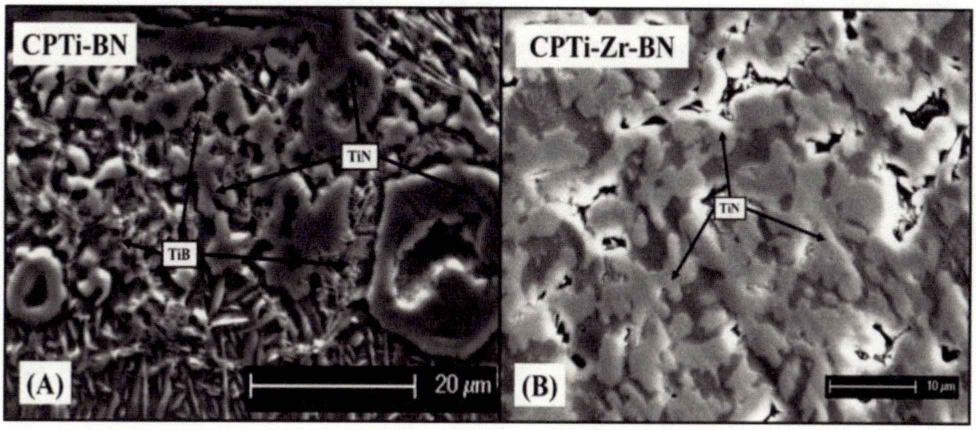

Figure 2.
SEM images of (A) CPTi-BN [45] (B) CPTi-Zr-BN [45] compositions.

through LENS technique. A specific single β-phase is identified in 20 wt% V while the dual-phase microstructure in Ti-12 wt% Mo system revealed α-phase in β-grain morphology. This was accredited to the build directions that had different thermal cycles at various portion of the builds.

In the work of Li et al. [47], LAM was used to fabricate Ti-6Al-2V-1.5Mo-0.5Zr-0.3Si alloy in order to examine tensile strength and microhardness of the alloy. Mixed columnar β-grains with no equiaxed microstructure was observed for as-deposited alloy. The addition of Zr, Mo and Si causes increase in microhardness value because of hardening effect of their solid solution. Also noticed, was the rise in ductility in comparison to Ti-6Al-4V.

Song et al. [5] reported the titanium alloy to fabricate a prosthetic beak as shown in (**Figure 3**) for a bird (*Grus japonensis*) using SLM. A model was first developed for the titanium alloy customized beak and was successfully completed to save the endangered bird.

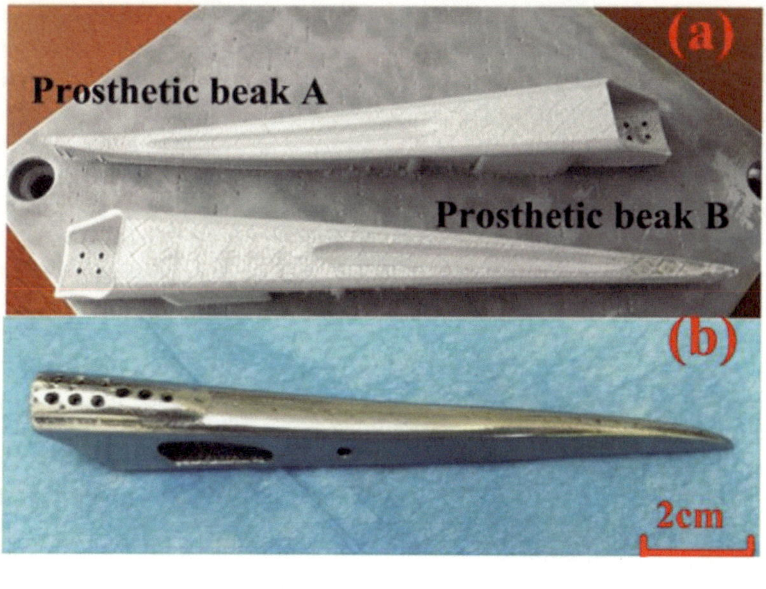

Figure 3.
SLM fabricated prosthesis beaks: (a) as-fabricated (b) polished and surface treated [5].

A comparative study of SLM and electron beam melting (EBM) carried out by Zhao et al. [54] to fabricate Ti-6Al-4V parts. Both components largely contained $\alpha + \beta$ and α' phase, respectively. SLM samples had higher tensile strength but lower ductility than EBM fabricated parts. Even though porosity was observed in both cases it was much higher in SLM parts. While generally, samples fabricated in the vertical direction had increased ultimate tensile strength, yield strength, and improve ductility than those in the horizontal orientation.

Direct metal deposition (DMD) has been used by Pouzet et al. [55] to build samples with varying volume fractions of TiB + TiC. The Coaxial deposition process fabricated TMC from powder mixture of T-6Al-4V + B4C. Pure and consistent Ti-6Al-4V-TiB microstructure was noticed and boron isolation during solidification of β-Ti resulted in grain refinements. At low B4C, TiC was not observed which was ascribed to high solubility of C in α-Ti at high solidification rate leading to decreased tensile strength at elevated and room temperature.

2.2 Laser additive manufacturing (LAM) of titanium aluminides

TiAl are known to be favorable for high-temperature component because of their high strength-to-weight ratio, high melting point, attractive creep and superior modulus of elasticity [56]. Recently, TiAl-based alloys have been effectively applied in aircraft engine productions with increase in demand for these alloys expected in the near future [57].

Microstructure and phase composition are immensely determining the properties of intermetallic TiAl-based alloys. Elements like Mo, V, Zr, Mn, Nb, Cr, Ni and Fe are known to be employed in developing enhanced mechanical properties for TiAl [58, 59]. The site effects for most of the elements remains unchanged with the different alloy compositions. However, Cr, Mn and V site substitution vary subject to Ti:Al ratio [59]. It is commonly accepted that a microstructure of equiaxed fine grains are desirable to enhance strength and ductility whereas coarse microstructure of fully lamellar is better for high creep.

The lack of ductility of TiAl at room temperature makes it problematic to machine and fabricate components by traditional methods. AM is very promising to incorporating design freedoms and processing when building components [60]. LENS and selective laser melting/sintering (SLM/SLS) are primarily known as LAM techniques. Various AM technologies are called by their trade names from the manufacturers or company of establishments [61]. A laser engineered net shaping (LENS) Optomec 850R machine for AM system is shown in **Figure 4**. This technique is suitable for manufacturing and realization of complex geometries with reduced lead time and material loss [62]. The process uses a focused laser beam for melting and fusion producing 3D parts from feedstock powder. Apart from building components (**Figure 5**), it refurbishes components through a process called laser cladding. This is believed to be the state-of-the-art technique for aero engine parts repairing [63].

For aerospace components materials, when compared with conventionally applied materials like Ni-based superalloys, TiAl is far harder and offers close to 50% weight savings. This results in decrease cost of turbine operations and emission levels due to highly efficient fuel usage [63]. Advances in recent development techniques has led to collaborations among manufacturers. This is made possible to fabricate turbine blades using electron beam melting (EBM) from pre-alloyed Ti4822 powder, presently employed in GEnx engines [60, 64].

For aerospace components materials, when compared with conventionally applied materials like Ni-based superalloys, TiAl far harder and offers close to 50% weight savings. This results in decrease cost of turbine operations and emission

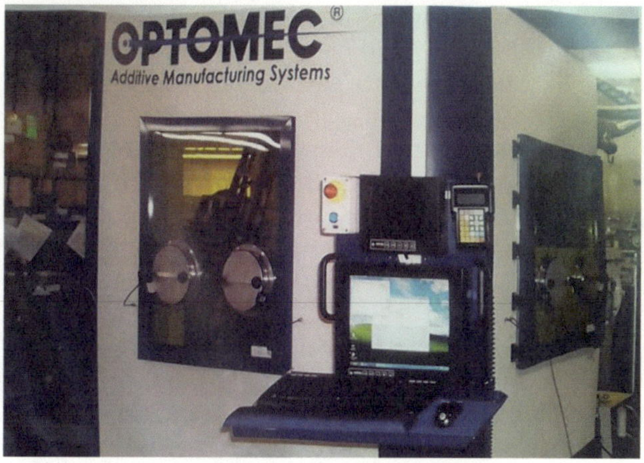

Figure 4.
LENS Optomec 850R additive manufacturing system.

Figure 5.
LENS Fabricated Object.

levels due to highly efficient fuel usage [63]. Advances in recent development techniques has led to collaborations among manufacturers. This is made possible to fabricate turbine blades using electron beam melting (EBM) from pre-alloyed Ti4822 powder, presently employed in GEnx engines [60, 64].

DED was adopted by Hoosain et al. [10] to create clads of γ-TiAl using in-situ elemental Ti and Al powder. The purpose of the study was to optimize formation of γ-phase during laser processing through mass flow rates. Depending on flow rate of Al and heat treatment both duplex and lamellar phases are observed. Also, Al content largely determines the microhardness and twinning existence. In another work, Hoosain et al. [56] investigated the consequences of Al flow rate on microstructure of in-situ alloy composition. The γ-TiAl was compared with Ti4822, a duplex phase microstructure was observed as Al feed rate increase. Fine lamellar grains were not generated for heat treatment temperature at 1200°C because it was below the temperature of α-transus. **Figure 6** presents the microstructure of commercial GE alloy Ti-48Al-2Cr-2Nb alloys.

In the feasibility study carried out by Dilip et al. [65], Ti-6Al-4V and Al powder was employed to samples of TiAl using binder jetting AM and reactive sintering.

Initially, Al_3Ti formation was obtained in the liquid phase due to Ti64 and Al reaction. Afterwards, TiAl was formed has diffusion of Al continues with sintering time. Other intermetallic together with TiAl phases were obtained in the final microstructure.

Todai et al. [41] presented a study to manipulate Ti4822 alloy microstructure fabricated through EBM process. Duplex-like and γ-bands of layered microstructure were formed by recurring heat treatment of melt pool area. The angle of building substantially makes the tensile elongation and yield strength (YS) to vary. The XRD patterns of the EBM built Ti-48Al-2Cr-2Nb alloy in different directions are displayed in **Figure 7**. The anisotropy of YS reduces as temperature rises. The heat treatment at 700°C gave the highest YS which show decrease in value as temperature rises. EBM was also adopted by Murr et al. [66] in fabricating TiAl from precursor powders to investigate to hardness and microstructure produced. Microstructure of α_2-Ti_3Al was noticed for powder precursor whereas lamellar colony at γ/α_2 with equiaxed γ-TiAl was obtained for EBM samples. It was reported that the EBM-fabricated TiAl had microhardness of ~4.1GPa which makes the YS far exceed that of EBM-fabricated Ti64.

Selective electron beam melting (SEBM) was adopted by Chen et al. [67], to investigate the microstructural variations of TiAl alloy samples and their impact on microhardness. A near lamellar structure was observed TiAl alloys. During the

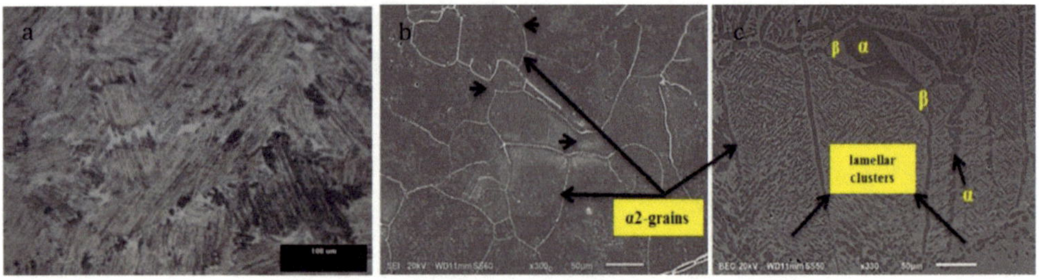

Figure 6.
Microstructure of the commercial GE alloy Ti-48Al-2Cr-2Nb (a) showing the lamellar type microstructure [56] and heat treated of at (b) 1200°C [10] and (c) 1430°C [10].

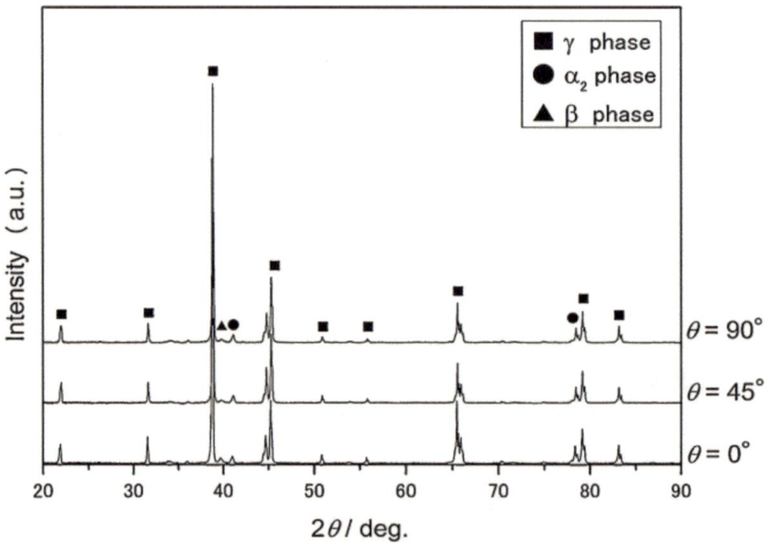

Figure 7.
XRD pattern of Ti-48Al-2Cr-2Nb alloy specimens fabricated by EBM at 0°, 45° and 90° [41].

process, three phase transformation of $\alpha_2 + \gamma \rightarrow B2$, $\alpha_2 \rightarrow B2$, $\alpha_2 \rightarrow \gamma$ was noticed. An increase in hardness was ascribed to rapidly raising quantity of lamellar colonies α_2/γ has the built height increases.

The structure of TiAl alloys was examined by Li et al. [68] for detailed characterization of SLM processed Ti-45Al-2Cr-5Nb alloy. Anisotropy of microstructure were obtained has the sides contained columnar grains while the top showed equiaxed microstructure. The processed sample was overshadowed with high angle grain boundaries (HAGBs) which is favored by the energy density. Also, evenly dispersed minute quantities of B2 and γ phases (nano meters range) were found in α_2 matrix γ and B2 phase reduces with raising energy density whereas α_2 content increases.

Sharman, Hughes and Ridgway [61] stated that previous works on LAM had concluded that it was merely impossible to manufacture component of TiAl that are crack free without additional heating to influence the solidification process. It was demonstrated that by defocusing the laser across build surface heats the particles ahead of entering melting zone. Hence, decrease rate of cooling lower than the critical level that permits crack formation.

Oxide dispersion strengthen (ODS) has been used by Kenel et al. [14] to fabricate Ti-45Al-3Nb- $< 0.2Y_2O_3$ at.% and studied the microstructure. The traditional microstructure of $\alpha_2 + \gamma$ and near-lamellar were observed. The ODS-TiAl alloy displays greater hardness than ones without the reinforcement. β-solidifying TiAl was studied by Gussone et al. [69] fabricated by SLM of γ-TiAl pre-alloyed powder. Under cooling of β occurred due to tremendous Al loss leading to β/B2 microstructure. During SLM and hot isostatic pressing (HIP), the raise in content of oxygen make the sample less ductile.

Seifi et al. [70] conducted and experiment to evaluate the mechanical properties and microstructure of EBM γ-TiAl with HIP. The two conditions presented microstructure of heterogenous interchanging γ-grains of coarse and fine bands. It is worthy to note that the EBM samples revealed microcracks at secluded sections all over the builds. The consequence of beam current on microstructure and grain boundary orientation had been studied by Yue et al. [71] for Ti-47Al-2Cr-2Nb manufactured by selective electron beam melting (SEBM). Increases beam current resulted in resulted in microstructural transformation to near-γ structure to fine duplex phase and B2-phase volume fraction steadily rises. Also, HAGB increases progressively with beam current while the low angle grain boundary (LAGB) reduces.

In establishing a relationship among scan speed, microstructure and mechanical properties, Li et al. [18] fabricated Ti-45Al-2Cr-2Nb parts using SLM. The grain size reduces with laser scan while crystallographic topography basically remains unchanged. The phases of B2 and γ rises with increased scan speed whereas α_2-phase is lowered. It was also reported that nanohardness and compressive strength rises with improved scan speed for the SLM TiAl alloy.

Chen, Yue and Wang [72] studied SEBM sample using various scan speed to produce Ti-47Al-2Cr-2Nb alloys. Microstructural changed occurred to duplex from near-γ structure as the scan speed rises. However, both α_2 and B2 phases are homogenously dispersed in the matrix of γ. Ultimate compressive strength (UTS) of SEBM-produced TiAl alloy generally increased with the scanning speed.

Statistical examination was used by Shi et al. [73] to optimize process parameter of SLM Ti-47Al-2Cr-2Nb on Ti-6Al-4V substrate. It was understood that unsuitable amalgamation of process parameters would lead to instability of melt pool, cracks balling and other defects. To optimize the process, Al loss need to be taken into account. The ideal process parameter obtained was able to build TiAl components with 97.34–98.95% densities.

3. Development of titanium aluminide composites for aerospace applications

LENS technique uses high-powered laser to deposit materials and it has been significant in repairing and building of metallic components [74]. Nevertheless, the qualities and mechanical properties of built components are greatly affected by imperfections like porosity, crack cavitation and inhomogeneous microstructure. Composite are of very high demand in aerospace industry because of requests for property combination like high stiffness, fracture toughness, strength, oxidation resistance and lightweight. Moreover, the impetus for composites in aerospace is the ever-increasing demand, thus substitute for metal alloys as weight reduction, cost-saving and fuel efficiency [75]. The aviation industry needs components made with materials with high thermal and mechanical properties at reduced cost [75, 76]. Presently, AM of TiAl is developing, but cracks do emanate due to the integrally brittle nature of the material [77].

According to Kablov et al. [78], the qualitative improvement of the technical and performance indicators of aviation gas turbine engines (AGTEs) and power plants largely depends on the implementation of new materials with a previously unachieved combination of properties. It was established that alloying intermetallic titanium ortho alloy with gadolinium in the amount of 0.1 at.% results to the rise in strength at room temperatures owing to the realization of the mechanism of dispersion strengthening with finely dispersed gadolinium oxides [78].

A crucible-less technique was adopted by Kartavykh et al. [79] to cast TiAl alloy with composition with composition of Ti-44Al-5Nb-3Cr-1.5Zr (at.%) and high-gradient float zone (FZ) re-solidification. Dual phase of $\gamma + \alpha_2$ lamellar with small quantities of B2 + γ was observed for the FZ processed alloy. This alloy had high tensile strength which is promoted by Cr accumulation in B2 phase.

A third-generation Ti-43.5Al-4Nb-1Mo-0.1B (TNM) fabricated with casting +HIP was studied by Dahar, Tamirisakandala and Lewandowski [27] to determine the influence of thermo mechanical treated specimen and applied force fracture behavior. Noteworthy alterations of phases were noticed due to the processing.

A coating of TiAl resistant to oxidation was developed by Sienkiewicz et al. [80], for thermal shield structures of material. A coating of titanium-aluminum-silicon system of varying chemical configurations was produced through warm spraying. Silicide of $Ti_5(Si,Al)_3$ was observed for two kind of morphologies. Generally, the coating shows attractive cyclic and isothermal resistance.

A complete processing chain develop by Juechter et al. [81], to demonstrated that Ti-45Al-4Nb-C processed by selective electron beam melting (SEBM) could be applied in manufacturing turbocharger wheels. This was made possible by enhanced optimization scan approach coupled with heat treatment.

The microstructures and mechanical properties of Ti6Al4V and Ti45Al7Nb powder mixtures in ratio 1:1 was studied by Wenjun, Chao and Feng [82] using Electron beam selective melting (EBSM) to build samples resulting in Ti22Al3.5Nb2V alloy. Two distinct portions of microstructure were noticed, a dual ($\alpha_2 + \beta$) phase and martensite portion. The dual phase region had lower microhardness value compared to the other portions.

The effects of carbon (C) addition was studied by Klein et al. [76], on γ-TiAl based alloys. Generally, C induces more α_2-phase formation and reduces β_o-phase. However, substantial hardening of γ and α_2 phases occurred with rise in C concentration.

Laser cladding TiN/Ti_3Al intermetallic composite coatings have been successfully used to increase the elevated temperature resistance to oxidation and mechanical properties of Ti6Al4V alloy [83]. The clads were largely constituted by phases of α-Ti, Ti_3Al and TiN. The intermetallic of TiN/Ti_3Al are high oxidation resistant with microhardness far above the value of Ti6Al4V.

An examination of the connection between boron (B) content and microstructural characteristics of TiAl/B (Ti-46.5Al-2.5Cr-2Nb-0.5Y/B) composites produced by SLM was carried out [84]. The grain size reduces as B increases but the HAGBs also rises with B. Owing to grain refinement strengthening system, the highest value of 1610.53 MPa and 5.17% was recorded for compressive strength and strain respectively.

The effects of Nb, Ta and Zr had been investigated by Bresler et al. [85], for Ti-44Al-5X (X = Nb, Ta, Zr) to examine creep properties (**Figure 8**). All alloying elements mostly improve creep performance substantially and reduce interspacing of α_2/γ while enhancing stability of microstructure.

TiAl-based alloys of two molybdenum (Mo) content was fabricated by Jiang et al. [86] through vacuum levitation melting (VLM). Mo addition is understood to refine lamellar size colonies. It also encourages twining capabilities and reduces cracks in β phase.

Laser metal deposition (LMD) was adopted by Carrullo, Falcón and Borrás [87], to analyzed films produced on Ti48Al2Cr2Nb coating during oxidation through optimization of parameters. The coatings demonstrated high oxidation resistance at 800°C. The layers of oxides formed showed complex microstructure and development of consecutive layers on coating.

Ambient temperature mechanical characteristics were evaluated by Dai et al. [88], for coatings of Ti-Al-Si on the substrate of Ti6Al4V through laser surface alloying. Thermal expansion between coating and substrate resulted in crack formation and propagation. The coating basically consists of Ti_5Si_3 reinforcement in Ti-Al-phase. Si content increase causes resultant rise in brittle Ti_5Si_3 which had increased hardness than Ti-Al. The process of high-temperature oxidation resistance of Si in Ti-Al-Si system coatings involves increased refining of oxide grains and boosting the creation of Al_2O_3.

An investigated of microstructure carried out by Maliutina et al. [89], on laser cladding of TA6Zr4DE using Ti48Al2Cr2Nb powders. The coatings had γ-TiAl and α_2-Ti_3Al phases of fully lamellar microstructure. The occurrence of niobium as an

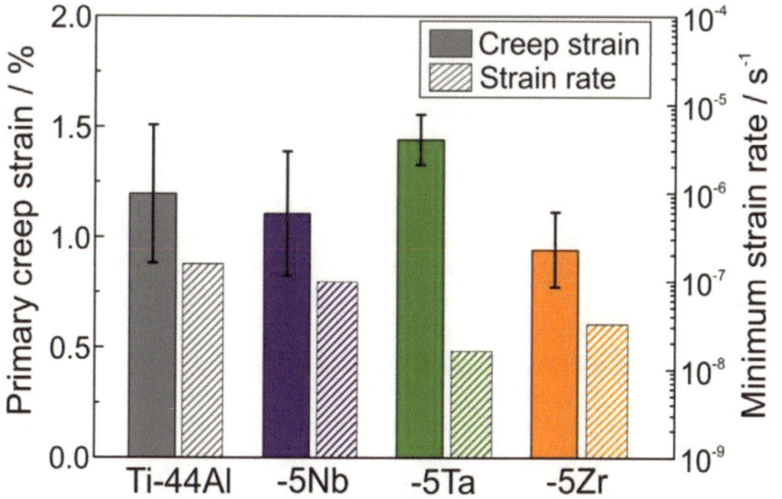

Figure 8.
Primary creep regime and minimum strain rate of binary Ti-44Al and Ti-44Al-5X (X = Nb, Ta, Zr) alloys, tested at 900°C and 100 MPa [85].

alloying element restrains the oxygen dispersion making impact in the creation of more favored alumina films.

LENS was employed by Zhang and Bandyopadhyay [90], to deposit Ti6Al4V and Al_2O_3 powders on Ti6Al4V substrate. For the pure Ti6Al4V, α-laths of Widmanstätten Ti was obtained while Ti6Al4V + Al_2O_3 parts revealed equiaxed grains with non-melted Al_2O_3. Microhardness results showed that Al_2O_3 section possessed the maximum hardness value followed by Ti6Al4V + Al_2O_3 sections.

Direct laser cladding (DLC) of Ti45Al5Nb0.5Si had been studied by Majumdar et al. [91], to determine the consequence of varying parameters on titanium aluminide. Dual phase $\alpha_2 + \gamma$ was revealed in DLC Ti45Al5Nb0.5Si alloy. The processing parameters had little effect on the microhardness of the clad. Rise in Si quantity improved the propensity to produce cracks.

4. Heat treatment of laser processed titanium aluminide components

The microstructural evolution of γ-TiAl clearly affects the end products mechanical properties as a result of processing and post-processing operations. Dependent on the characteristics anticipated for parts, several techniques could be utilized to fix microstructure. It has been reported that lamellar of coarse grains display comparatively superior fracture toughness, creep but lack ductility, notably at room temperature whilst duplex of fine grains connotes low fracture toughness, creep resistance nonetheless modest ductile property [27].

SLM of Ti-44.8Al-6Nb-1.0Mo-0.1B was investigated by Gussone et al. [62] and influence of heat-treatment during and after fabrication. The results showed that considerable changes stimulated by heating the unstable Ti-44.8Al-6Nb-1.0Mo-0.1B varies sharply. The initial heat treating of samples produces microstructures exemplified by increased orbicular γ and β/βo at the α_2/γ boundaries colony.

Alloy Ti-47Al-2Nb-2Cr (at%) fabricated through selective electron beam melting (SEBM) by Guangyu et al. [92], examined the influence of post processing heat treatment on microstructure. Lamellae microstructure was revealed for the samples (**Figure 9**) because of SEBM cyclic heating. Hence, transformation of phases was clearly through β-phase. Though, various structures are gotten at different temperatures, oil quenching at 1250°C+ heating to 1200°C for 2 h produces a refined and homogeneous microstructure.

Klein et al. [93] investigated the microstructural formations of Ti-Al-Nb-Cr-Mo in understanding the requirements for controlling the microstructure formation and concomitant properties. Upon annealing, γ-platelets arise initially from the precipitating β_o-phase. Cr and Mo are repelled throughout the γ-grain development.

The influence of annealing temperature was investigated by Ren et al. [19], on Ti-45Al-8.5Nb formation of controlled in lamellar structure. The regulated α_2 lamellae appears in alloys annealed at lesser than 650°C and vanishes beyond 700°C. Extensive hardening stimulates O phase on α_2 lamellae and the $\gamma/(\alpha_2 + O)$ been obtained attributed to refined α_2 lamellae. The schematic of the heat treatment procedure is depicted in **Figure 10** below.

Analysis of microstructure of high-Nb TiAl alloy carried out by Qiang et al. [94], for Ti-45Al-8.5Nb-0.2 W-0.2B-0.02Y (at.%) resulting from numerous cooling rate. The transformation path $\alpha \rightarrow \alpha_2 + \gamma$ for lamellar microstructure is controlled chiefly by its rate of cooling. Fast rate of cooling causes the retention of α/α_2 phase of supersaturated microstructure. It was suggested that a rapid cooling rate and slow rate of cooling for $\alpha \rightarrow \alpha_2 + \gamma$ and $\beta \rightarrow \alpha$ transformation respectively.

In the repair of defective or worn blade parts by Rittinghaus et al. [95], LMD of TNM™ β-containing alloy with composition of Ti-43.5Al-4Nb-1Mo-0.1B (at.%) was

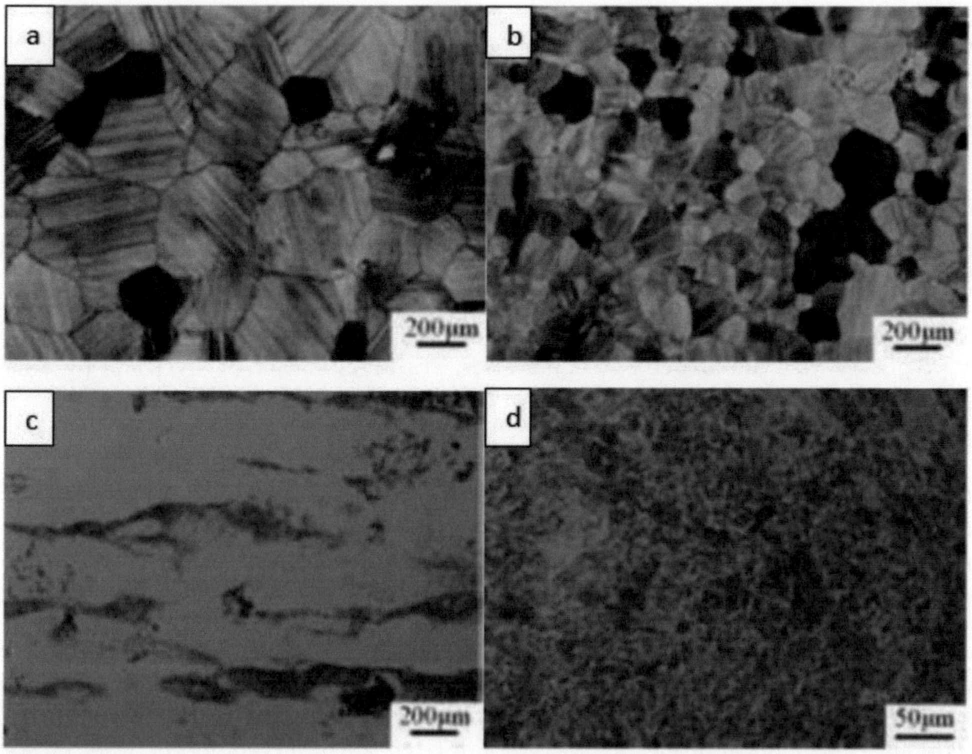

Figure 9.
Microstructures of heat-treated Ti-47Al-2Nb-2Cr alloy at (a) 1250°C for 2 h air cooled (b) 1300°C for 10 min air cooled (c) 1250°C for 10 min oil quenched (d) 1250°C for 10 min oil quenched followed heat treating at 1200°C for 2 h air cooled [92].

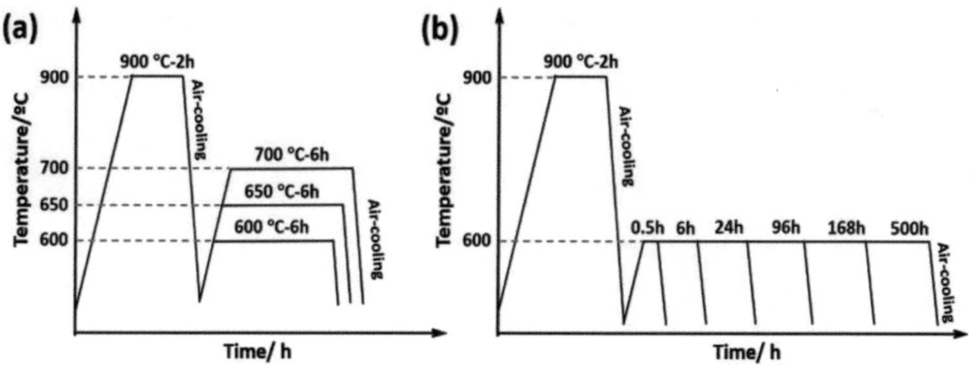

Figure 10.
Heat Treatment Pattern Adopted by Ren et al. 19 for the γ-Ti-45Al-8.5Nb alloy annealed at 900 oC for 2 hrs followed by air cooling with (a) annealing at 600 °C, 650 °C and 700 °C for 6 hrs each followed by air cooling (b) isothermal annealing at 600 °C for annealing time range from 0.5 to 500 hrs followed by air cooling.

studied. Small fraction of β_o reduces the microhardness of the LMD alloy. Lamellar orderings are moderately greater in the LMD microstructure because annealing is achieved in the $(\gamma + \alpha)$ phase with noticeable coarsening of the non-melted γ lamellae.

Also, phase transformations were examined by Kastenhuber et al. [28] for β-containing γ-TiAl based alloy in gas atomization of fast cooling. Initial variations of composition caused a primary β-phase, indicating to an uninterrupted escalation of the Al and lessening of Nb, Mo and Ti at the previous β-grain boundaries. Therefore, following $L \rightarrow L + \beta$ reaction the solidification path continues unevenly, subject to Al-content and quantities of Nb and/or Mo. At room temperature the γ-phase appears in three different morphologies.

The consequence of heat treatments was studied by Li et al. [96], to evaluate grains features, transformation of phases and hardness of Ti-45Al-2Cr-5Nb (at.%) produced through SLM. The grains of heat-treated parts are overshadowed with HAGBs that rises as temperature of annealing increases. SLM-fabricated alloy has $D0_{19}$ structure of α_2 phase but the B2-phase is a bcc structure and γ-phase illustrate a $L1_o$ (bcc) structure.

The microstructural evolution and room temperature fracture toughness β-solidifying TiAl alloy studied by Chen et al. [30], for Ti-45Al-2Nb-1.5V-1Mo-0.3Y (at.%). Fine near-lamellar containing majorly of fine grains and combinations of γ + β phases with lamellar colony boundaries. The precipitation of fine β and γ grains is deemed an inherent toughness system due to α_2/β and α_2/γ boundaries creating limited displacement movement successfully.

The influence of heat treatment was studied by Tebaldo and Faga [97], on hardness of Ti48Al2Nb2Cr manufactured through EBM technology. It was noticed that only minor different in density occurred due heat treatment and micro porosity. Near-γ with equiaxed lamellar was obtained for samples not heat treated and fully lamellar achieved for heat treated samples. It was concluded that the developed alloy would be appropriate for manufacturing components in aircraft engines.

An incorporation of both HIP and heat treatment was developed by Chen et al. [13], to enhance mechanical properties of TiAl alloy. Corresponding increment in elongation and YS was noticed for both duplex and lamellar structures. Reduction in TiAl alloy mechanical properties was evident from the observed microcracks. However, the combined production method showed a microcrack free microstructure, thereby enhancing the properties. Illustration of the processing routes are shown in **Figure 11**.

Mechanical and microstructure governing techniques was investigated by Zhao et al. [98], for Ti-46Al-2Nb-2V-1Mo-Y (at%). The α + β + γ phase pre-treatment by annealing actually gave raise to the breakdown of lamellar α/γ structures. Fully-lamellar was attained with prolong annealing whereas for two-step process,

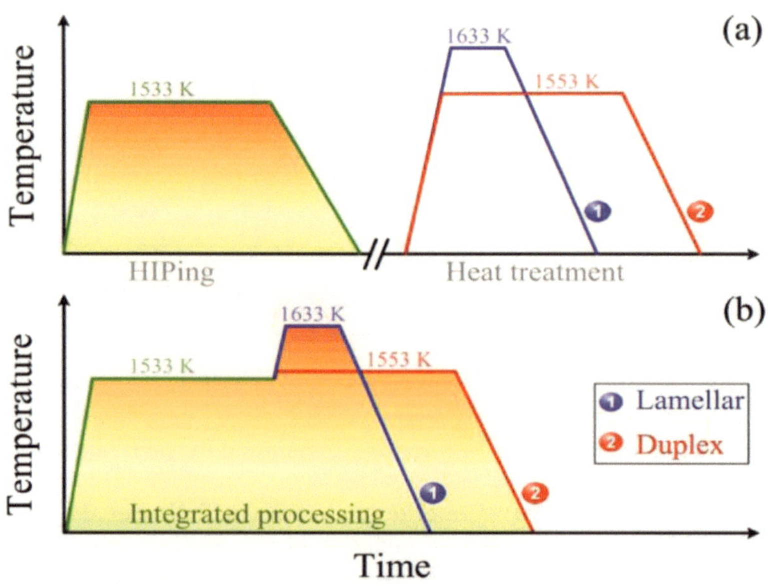

Figure 11.
The processing routes adopted by Chen et al. [13] (a) separately (b) combined processing.

duplex phase were observed. The duplex phase reveals greater room temperature tensile property, whilst fully-lamellar had better strength.

The influence of cooling rate was investigated by Fan et al. [99], on mechanical and microstructural attributes of Ti-49Al alloy. Increased rate of cooling results in reduction of interlamellar and dendrite spacings. Whereas rise in cooling rate causes corresponding improvement in tensile strength and hardness values. Consequently, microhardness value rises as tensile strength increases for Ti-49Al (at.%) directionally solidified alloy.

5. Challenges and critical issues

According to Bünck et al. [58], recent manufacturing processes for aerospace parts are expected to be well developed and cost-effectively competitive methods. LAM parts have high surface roughness and structural inhomogeneity and usually necessitate supplementary treatments to diminish irregular microstructure. Inadequate ambient temperature ductile property of TiAl alloys deters processing. Even during forging using isothermal process, dies are preserved at considerably high temperatures. Practical issue still to be stopped is the reactivity of melting crucible with TiAl, besides the matters of cost [25]. Novel methods are required that permit the modification of satisfactory amount of ductility though preserving the exceptional strength and creep characteristics that aforementioned research works has unlocked. The solution to unravel this difficulty rests in essential consideration of fundamental relationship between the properties and microstructure mechanisms. Grain morphology control is a challenging issue for LAM of large metallic components. Numerous practices have extensively considered the additional characteristics of TiAl-based aerospace parts, comprising the practice of surface modeling, application of coatings and heat-treatment.

Presently, LAM has the prospective to develop typical manufacture techniques, specifically in creating of small size of extremely complicated components [100]. Manufacturers have previously presented low-pressure turbine blades (LPTB) of TiAl alloy in aero engines and more interest is shown for the development of TiAl, demonstrating the distinctiveness of the alloy. This is with a view that engine efficiency would be substantially better-quality, indicating decreased fuel utilization, CO_2 emission and noise. Consequently, increases stress is envisaged for TiAl-LPTBs and necessitates adjusting processing routes to produce functional materials.

LAM is suggested to have benefits for aircraft engine components because it can easily translate designs to 3D component, produce customized parts and functional designs with complex and intricate features [36]. It also has potential of manufacturing parts with no waste, lightweight and minimal lead time of manufacturing. Hence, it has the ability to be exceptional scalable and manufacture final product with little or no post-processing.

6. Summary and conclusions

Based on the reviewed works, it can be generally deduced that LAM processing of titanium aluminide suffers several set-backs militating against its wider acceptance in aerospace application. Although, it has attractive features making it to be qualified as a candidate material for aerospace applications, the processing and post-processing problems still needs to be looked into. It is expected that combined processes of alloying and heat treatment (in-situ and ex-situ) could result in appreciable ductility while maintaining other mechanical properties. Consequently, researchers would be able to develop a mechanism of fabricating TiAl with better room and elevated temperature properties.

Acknowledgements

The authors would like to acknowledge the financial support of African Laser Centre-National Laser Centre; Council of Scientific and Industrial Research (ALC-NLC; CSIR), Project Number LHIP500 Task ALC S100.

Author details

Sadiq Abiola Raji[1*], Abimbola Patricia Idowu Popoola[1], Sisa Leslie Pityana[1,2], Olawale Muhmmed Popoola[3,4], Fatai Olufemi Aramide[1,5], Monnamme Tlotleng[2] and Nana Kwamina Kum Arthur[2]

1 Department of Chemical, Metallurgical and Materials Engineering, Tshwane University of Technology, Pretoria, South Africa

2 National Laser Centre, Council of Scientific and Industrial Research (NLC-CSIR), Pretoria, South Africa

3 Department of Electrical Engineering, Tshwane University of Technology, Pretoria, South Africa

4 Centre for Energy and Electric Power (CEEP), Tshwane University of Technology, Pretoria, South Africa

5 Department of Metallurgical and Materials Engineering, Federal University of Technology, Akure, Ondo State, Nigeria

*Address all correspondence to: rajchandy2355@gmail.com

References

[1] Zhang M, Shen M, Xin L, Ding X, Zhu S, Wang F. High vacuum arc ion plating TiAl coatings for protecting titanium alloy against oxidation at medium high temperatures. Corrosion Science. 2016;**112**:36-43. DOI: 10.1016/j.corsci.2016.07.005

[2] Sallica-Leva E, Jardini AL, Fogagnolo JB. Microstructure and mechanical behavior of porous Ti–6Al–4V parts obtained by selective laser melting. Journal of the Mechanical Behavior of Biomedical Materials. 2013;**26**:98-108. DOI: 10.1016/j.jmbbm.2013.05.011

[3] Hu Y, Ning F, Wang H, Cong W, Zhao B. Laser engineered net shaping of quasi-continuous network microstructural TiB reinforced titanium matrix bulk composites: Microstructure and wear performance. Optics & Laser Technology. 2018;**99**:174-183. DOI: 10.1016/j.optlastec.2017.08.032

[4] Mohammad A, Al-Ahmari AM, Balla VK, Das M, Datta S, Yadav D, et al. In vitro wear, corrosion and biocompatibility of electron beam melted γ-TiAl. Materials & Design. 2017;**133**:186-194. DOI: 10.1016/j.matdes.2017.07.065

[5] Song C, Wang A, Wu Z, Chen Z, Yang Y, Wang D. The design and manufacturing of a titanium alloy beak for Grus japonensis using additive manufacturing. Materials & Design. 2017;**117**:410-416. DOI: 10.1016/j.matdes.2016.11.092

[6] Hu Y, Zhao B, Ning F, Wang H, Cong W. In-situ ultrafine three-dimensional quasi-continuous network microstructural TiB reinforced titanium matrix composites fabrication using laser engineered net shaping. Materials Letters. 2017;**195**:116-119. DOI: 10.1016/j.matlet.2017.02.112

[7] Liang YJ, Liu D, Wang HM. Microstructure and mechanical behavior of commercial purity Ti/Ti–6Al–2Zr–1Mo–1V structurally graded material fabricated by laser additive manufacturing. Scripta Materialia. 2014;**74**:80-83. DOI: 10.1016/j.scriptamat.2013.11.002

[8] Santhosh R, Geetha M, Rao MN. Recent developments in heat treatment of beta titanium alloys for aerospace applications. Transactions of the Indian Institute of Metals. 2017;**70**(7):1681-1688. DOI: 10.1007/s12666-016-0985-6

[9] Tlotleng M, Lengopeng T, Seerane MN, Pityana SL. Effects of heat-treatment on the microstructure of TiAl-Nb produced with laser metal deposition technique. In: Proceedings of the Materials Science & Technology, October 8–12, 2017. Pittsburgh, Pennsylvania USA: David L. Lawrence Convention Center; 2017. DOI: http://hdl.handle.net/10204/9874

[10] Hoosain S, Pityana S, Freemantle C, Tlotleng M. Heat treatment of in situ laser-fabricated titanium aluminide. Metals. 2018;**8**(9):655. DOI: 10.3390/met8090655

[11] Kartavykh AV, Asnis EA, Piskun NV, Statkevich II, Gorshenkov MV, Korotitskiy AV. A promising microstructure/deformability adjustment of β-stabilized γ-TiAl intermetallics. Materials Letters. 2016;**162**:180-184. DOI: 10.1016/j.matlet.2015.09.139

[12] Cheng J, Yu Y, Fu L, Li F, Qiao Z, Li J, et al. Effect of TiB2 on dry-sliding tribological properties of TiAl intermetallics. Tribology International. 2013;**62**:91-99. DOI: 10.1016/j.triboint.2013.02.006

[13] Chen L, Zhu L, Guan Y, Zhang B, Li J. Tougher TiAl alloy via integration of hot isostatic pressing and heat treatment. Materials Science and

Engineering A. 2017;**688**:371-377. DOI: 10.1016/j.msea.2017.02.028

[14] Kenel C, Dawson K, Barras J, Hauser C, Dasargyri G, Bauer T, et al. Microstructure and oxide particle stability in a novel ODS γ-TiAl alloy processed by spark plasma sintering and laser additive manufacturing. Intermetallics. 2017;**90**:63-73. DOI: 10.1016/j.intermet.2017.07.004

[15] Xu WC, Huang K, Wu SF, Zong YY, Shan DB. Influence of Mo content on microstructure and mechanical properties of β-containing TiAl alloy. Transactions of Nonferrous Metals Society of China. 2017;**27**(4):820-828. DOI: 10.1016/S1003-6326(17)60094-3

[16] Cho K, Kobayashi R, Oh JY, Yasuda HY, Todai M, Nakano T, et al. Influence of unique layered microstructure on fatigue properties of Ti-48Al-2Cr-2Nb alloys fabricated by electron beam melting. Intermetallics. 2018;**95**:1-10. DOI: 10.1016/j.intermet.2018.01.009

[17] Erdely P, Staron P, Maawad E, Schell N, Clemens H, Mayer S. Lattice and phase strain evolution during tensile loading of an intermetallic, multi-phase γ-TiAl based alloy. Acta Materialia. 2018;**158**:193-205. DOI: 10.1016/j.actamat.2018.07.062

[18] Li W, Liu J, Zhou Y, Li S, Wen S, Wei Q, et al. Effect of laser scanning speed on a Ti-45Al-2Cr-5Nb alloy processed by selective laser melting: Microstructure, phase and mechanical properties. Journal of Alloys and Compounds. 2016;**688**:626-636. DOI: 10.1016/j.jallcom.2016.07.206

[19] Ren GD, Dai CR, Mei W, Sun J, Lu S, Vitos L. Formation and temporal evolution of modulated structure in high Nb-containing lamellar γ-TiAl alloy. Acta Materialia. 2019;**165**:215-227. DOI: 10.1016/j.actamat.2018.11.041

[20] Akhlaghi M, Tayebifard SA, Salahi E, Asl MS. Spark plasma sintering of TiAl–Ti3AlC2 composite. Ceramics International. 2018;**44**(17):21759-21764. DOI: 10.1016/j.ceramint.2018.08.272

[21] Bakan E, Mauer G, Sohn YJ, Schwedt A, Rackel MW, Riedlberger F, et al. Cold gas spraying of Ti-48Al-2Cr-2Nb intermetallic for jet engine applications. Surface and Coatings Technology. 2018. In Press. DOI: 10.1016/j.surfcoat.2018.11.092

[22] Baudana G, Biamino S, Klöden B, Kirchner A, Weißgärber T, Kieback B, et al. Electron beam melting of Ti-48Al-2Nb-0.7 Cr-0.3 Si: Feasibility investigation. Intermetallics. 2016;**73**: 43-49. DOI: 10.1016/j.intermet.2016.03.001

[23] Kastenhuber M, Rashkova B, Clemens H, Mayer S. Effect of microstructural instability on the creep resistance of an advanced intermetallic γ-TiAl based alloy. Intermetallics. 2017; **80**:1-9. DOI: 10.1016/j.intermet.2016.09.007

[24] Dahar MS, Tamirisakandala SA, Lewandowski JJ. Fatigue crack growth and fracture behavior of as-cast Ti-43.5 Al-4Nb-1Mo-0.1 B (TNM) compared to Ti-48Al-2Nb-2Cr (4822). Intermetallics. 2017;**91**:158-168. DOI: 10.1016/j.intermet.2017.08.015

[25] Mohammad A, Alahmari AM, Mohammed MK, Renganayagalu RK, Moiduddin K. Effect of energy input on microstructure and mechanical properties of titanium aluminide alloy fabricated by the additive manufacturing process of electron beam melting. Materials. 2017;**10**(2):211. DOI: 10.3390/ma10020211

[26] Tang HP, Yang GY, Jia WP, He WW, Lu SL, Qian M. Additive manufacturing of a high niobium-containing titanium aluminide alloy by selective electron beam melting. Materials Science and Engineering A. 2015;**636**:103-107. DOI: 10.1016/j.msea.2015.03.079

[27] Dahar MS, Tamirisakandala SA, Lewandowski JJ. Evolution of fatigue crack growth and fracture behavior in gamma titanium aluminide Ti-43.5 Al-4Nb-1Mo-0.1 B (TNM) forgings. International Journal of Fatigue. 2018; **111**:54-69. DOI: 10.1016/j.ijfatigue.2018.01.026

[28] Kastenhuber M, Klein T, Rashkova B, Weißensteiner I, Clemens H, Mayer S. Phase transformations in a β-solidifying γ-TiAl based alloy during rapid solidification. Intermetallics. 2017; **91**:100-109. DOI: 10.1016/j.intermet.2017.08.017

[29] Fang H, Chen R, Chen X, Yang Y, Su Y, Ding H, et al. Effect of Ta element on microstructure formation and mechanical properties of high-Nb TiAl alloys. Intermetallics. 2019;**104**:43-51. DOI: 10.1016/j.intermet.2018.10.017

[30] Chen Y, Niu H, Kong F, Xiao S. Microstructure and fracture toughness of a β phase containing TiAl alloy. Intermetallics. 2011;**19**(10):1405-1410. DOI: 10.1016/j.intermet.2011.05.006

[31] Saboori A, Gallo D, Biamino S, Fino P, Lombardi M. An overview of additive manufacturing of titanium components by directed energy deposition: Microstructure and mechanical properties. Applied Sciences. 2017;7(9): 883. DOI: 10.3390/app7090883

[32] Calignano F, Manfredi D, Ambrosio EP, Biamino S, Lombardi M, Atzeni E, et al. Overview on additive manufacturing technologies. Proceedings of the IEEE. 2017;**105**(4): 593-612. DOI: 10.1109/JPROC.2016.2625098

[33] Hu Y, Ning F, Cong W, Li Y, Wang X, Wang H. Ultrasonic vibration-assisted laser engineering net shaping of ZrO2-Al2O3 bulk parts: Effects on crack suppression, microstructure, and mechanical properties. Ceramics International. 2018;**44**(3):2752-2760. DOI: 10.1016/j.ceramint.2017.11.013

[34] Onuike B, Heer B, Bandyopadhyay A. Additive manufacturing of Inconel 718—Copper alloy bimetallic structure using laser engineered net shaping (LENS™). Additive Manufacturing. 2018;**21**:133-140. DOI: 10.1016/j.addma.2018.02.007

[35] Yadollahi A, Shamsaei N. Additive manufacturing of fatigue resistant materials: Challenges and opportunities. International Journal of Fatigue. 2017; **98**:14-31. DOI: 10.1016/j.ijfatigue.2017.01.001

[36] Tofail SA, Koumoulos EP, Bandyopadhyay A, Bose S, O'Donoghue L, Charitidis C. Additive manufacturing: Scientific and technological challenges, market uptake and opportunities. Materials Today. 2018;**21**(1):22-37. DOI: 10.1016/j.mattod.2017.07.001

[37] Xu W, Lui EW, Pateras A, Qian M, Brandt M. In situ tailoring microstructure in additively manufactured Ti-6Al-4V for superior mechanical performance. Acta Materialia. 2017;**125**:390-400. DOI: 10.1016/j.actamat.2016.12.027

[38] Bourell D, Kruth JP, Leu M, Levy G, Rosen D, Beese AM, et al. Materials for additive manufacturing. CIRP Annals. 2017;**66**(2):659-681. DOI: 10.1016/j.cirp.2017.05.009

[39] Dutta B, Froes FS. The additive manufacturing (AM) of titanium alloys. Metal Powder Report. 2017;**72**(2): 96-106. DOI: 10.1016/j.mprp.2016.12.062

[40] Yan S, Wu D, Huang Y, Liu N, Zhang Y, Niu F, et al. C fiber toughening Al_2O_3-ZrO_2 eutectic via ultrasonic-assisted directed laser deposition. Materials Letters. 2019;**235**: 228-231. DOI: 10.1016/j.matlet.2018.10.047

[41] Todai M, Nakano T, Liu T, Yasuda HY, Hagihara K, Cho K, et al. Effect of building direction on the microstructure

and tensile properties of Ti-48Al-2Cr-2Nb alloy additively manufactured by electron beam melting. Additive Manufacturing. 2017;**13**:61-70. DOI: 10.1016/j.addma.2016.11.001

[42] Hooper PA. Melt pool temperature and cooling rates in laser powder bed fusion. Additive Manufacturing. 2018; **22**:548-559. DOI: 10.1016/j.addma.2018.05.032

[43] Attar H, Ehtemam-Haghighi S, Kent D, Dargusch MS. Recent developments and opportunities in additive manufacturing of titanium-based matrix composites: A review. International Journal of Machine Tools and Manufacture. 2018;**133**:85-102. DOI: 10.1016/j.ijmachtools.2018.06.003

[44] Santos LS, Gupta SK, Bruck HA. Simulation of buckling of internal features during selective laser sintering of metals. Additive Manufacturing. 2018;**23**:235-245. DOI: 10.1016/j.addma.2018.08.002

[45] Traxel KD, Bandyopadhyay A. Reactive-deposition-based additive manufacturing of Ti-Zr-BN composites. Additive Manufacturing. 2018;**24**: 353-363. DOI: 10.1016/j.addma.2018.10.005

[46] Zhu YY, Tang HB, Li Z, Xu C, He B. Solidification behavior and grain morphology of laser additive manufacturing titanium alloys. Journal of Alloys and Compounds. 2019;**777**: 712-716. DOI: 10.1016/j.jallcom.2018.11.055

[47] Li GC, Li J, Tian XJ, Cheng X, He B, Wang HM. Microstructure and properties of a novel titanium alloy Ti-6Al-2V-1.5 Mo-0.5 Zr-0.3 Si manufactured by laser additive manufacturing. Materials Science and Engineering A. 2017;**684**:233-238. DOI: 10.1016/j.msea.2016.11.084

[48] Mantri SA, Banerjee R. Microstructure and micro-texture evolution of additively manufactured β-Ti alloys. Additive Manufacturing. 2018;**23**:86-98. DOI: 10.1016/j.addma.2018.07.013

[49] Tang H, Tao W, Wang H, Song Y, Jian X, Yin L, et al. High-performance infrared emissivity of micro-arc oxidation coatings formed on titanium alloy for aerospace applications. International Journal of Applied Ceramic Technology. 2018;**15**(3): 579-591. DOI: 10.1111/ijac.12861

[50] Grove T, Denkena B, Maiß O, Krödel A, Schwab H, Kühn U. Cutting mechanism and surface integrity in milling of Ti-5553 processed by selective laser melting. Journal of Mechanical Science and Technology. 2018;**32**(10): 4883-4892. DOI: 10.1007/s12206-018-0936-8

[51] Ning F, Hu Y, Cong W. Microstructure and mechanical property of TiB reinforced Ti matrix composites fabricated by ultrasonic vibration-assisted laser engineered net shaping. Rapid Prototyping Journal. 2018;**25**(3):581-591. DOI: 10.1108/RPJ-05-2018-0118

[52] Uhlmann E, Kersting R, Klein TB, Cruz MF, Borille AV. Additive manufacturing of titanium alloy for aircraft components. Procedia CIRP. 2015;**35**:55-60. DOI: 10.1016/j.procir.2015.08.061

[53] Heer B, Bandyopadhyay A. Silica coated titanium using laser engineered net shaping for enhanced wear resistance. Additive Manufacturing. 2018;**23**:303-311. DOI: 10.1016/j.addma.2018.08.022

[54] Zhao X, Li S, Zhang M, Liu Y, Sercombe TB, Wang S, et al. Comparison of the microstructures and mechanical properties of Ti–6Al–4V fabricated by selective laser melting and electron beam melting. Materials & Design. 2016;**95**:21-31. DOI: 10.1016/j.matdes.2015.12.135

[55] Pouzet S, Peyre P, Gorny C, Castelnau O, Baudin T, Brisset F, et al. Additive layer manufacturing of titanium matrix composites using the direct metal deposition laser process. Materials Science and Engineering A. 2016;**677**:171-181. DOI: 10.1016/j.msea.2016.09.002

[56] Hoosain SE, Pityana SL, Tlotleng M, Legopeng T. A comparative study on laser processing of commercially available titanium aluminide (TI-48Al-2Cr-2Nb) and in-situ alloying of titanium aluminide. In: Proceedings of the 18th Annual International Rapid Product Development Association of South Africa (RAPDASA); 7–10 November 2017, Durban ICC, RAPDASA; South Africa 2017. DOI: hdl.handle.net/10204/9805

[57] Bünck M, Stoyanov T, Schievenbusch J, Michels H, Gußfeld A. Titanium aluminide casting technology development. JOM. 2017;**69**(12):2565-2570. DOI: 10.1007/s11837-017-2534-0

[58] Imayev RM, Imayev VM, Oehring M, Appel F. Alloy design concepts for refined gamma titanium aluminide based alloys. Intermetallics. 2007;**15**(4):451-460. DOI: 10.1016/j.intermet.2006.05.003

[59] Doubenskaia M, Domashenkov A, Smurov I, Petrovskiy P. Study of selective laser melting of intermetallic TiAl powder using integral analysis. International Journal of Machine Tools and Manufacture. 2018;**129**:1-4. DOI: 10.1016/j.ijmachtools.2018.02.003

[60] Catchpole-Smith S, Clare AT. In-situ synthesis of titanium aluminides by direct metal deposition. Journal of Materials Processing Technology. 2017;**239**:230-239. DOI: 10.1016/j.jmatprotec.2016.08.031

[61] Sharman AR, Hughes JI, Ridgway K. Characterisation of titanium aluminide components manufactured by laser metal deposition. Intermetallics. 2018;**93**:89-92. DOI: 10.1016/j.intermet.2017.11.013

[62] Gussone J, Garces G, Haubrich J, Stark A, Hagedorn YC, Schell N, et al. Microstructure stability of γ-TiAl produced by selective laser melting. Scripta Materialia. 2017;**130**:110-113. DOI: 10.1016/j.scriptamat.2016.11.028

[63] Kimme T, Seifert M. Laser surface cladding of titanium aluminides: Technology and engineering for the turbine blade repair. Laser Technik Journal. 2017;**14**(5):18-20. DOI: 10.1002/latj.201700026

[64] Bacos MP, Thomas M, Raviart JL, Morel A, Mercier S, Josso P. Influence of an oxidation protective coating upon hot corrosion and mechanical behaviour of Ti–48Al–2Cr–2Nb alloy. Intermetallics. 2011;**19**(8):1120-1129. DOI: 10.1016/j.intermet.2011.03.014

[65] Dilip JJ, Miyanaji H, Lassell A, Starr TL, Stucker B. A novel method to fabricate TiAl intermetallic alloy 3D parts using additive manufacturing. Defence Technology. 2017;**13**(2):72-76. DOI: 10.1016/j.dt.2016.08.001

[66] Murr LE, Gaytan SM, Ceylan A, Martinez E, Martinez JL, Hernandez DH, et al. Characterization of titanium aluminide alloy components fabricated by additive manufacturing using electron beam melting. Acta Materialia. 2010;**58**(5):1887-1894. DOI: 10.1016/j.actamat.2009.11.032

[67] Chen Y, Yue H, Wang X, Xiao S, Kong F, Cheng X, et al. Selective electron beam melting of TiAl alloy: Microstructure evolution, phase transformation and microhardness. Materials Characterization. 2018;**142**:584-592. DOI: 10.1016/j.matchar.2018.06.027

[68] Li W, Liu J, Wen S, Wei Q, Yan C, Shi Y. Crystal orientation, crystallographic texture and phase

evolution in the Ti–45Al–2Cr–5Nb alloy processed by selective laser melting. Materials Characterization. 2016;**113**: 125-133. DOI: 10.1016/j.matchar.2016.01.012

[69] Gussone J, Hagedorn YC, Gherekhloo H, Kasperovich G, Merzouk T, Hausmann J. Microstructure of γ-titanium aluminide processed by selective laser melting at elevated temperatures. Intermetallics. 2015;**66**: 133-140. DOI: 10.1016/j.intermet.2015.07.005

[70] Seifi M, Salem AA, Satko DP, Ackelid U, Semiatin SL, Lewandowski JJ. Effects of HIP on microstructural heterogeneity, defect distribution and mechanical properties of additively manufactured EBM Ti-48Al-2Cr-2Nb. Journal of Alloys and Compounds. 2017; **729**:1118-1135. DOI: 10.1016/j.jallcom.2017.09.163

[71] Yue H, Chen Y, Wang X, Kong F. Effect of beam current on microstructure, phase, grain characteristic and mechanical properties of Ti-47Al-2Cr-2Nb alloy fabricated by selective electron beam melting. Journal of Alloys and Compounds. 2018;**750**: 617-625. DOI: 10.1016/j.jallcom.2018.03.343

[72] Chen Y, Yue H, Wang X. Microstructure, texture and tensile property as a function of scanning speed of Ti-47Al-2Cr-2Nb alloy fabricated by selective electron beam melting. Materials Science and Engineering A. 2018;**713**:195-205. DOI: 10.1016/j.msea.2017.12.020

[73] Shi X, Ma S, Liu C, Wu Q. Parameter optimization for Ti-47Al-2Cr-2Nb in selective laser melting based on geometric characteristics of single scan tracks. Optics & Laser Technology. 2017;**90**:71-79. DOI: 10.1016/j.optlastec.2016.11.002

[74] Cong W, Ning F. A fundamental investigation on ultrasonic vibration-assisted laser engineered net shaping of stainless steel. International Journal of Machine Tools and Manufacture. 2017; **121**:61-69. DOI: 10.1016/j.ijmachtools.2017.04.008

[75] Toozandehjani M, Kamarudin N, Dashtizadeh Z, Lim EY, Gomes A, Gomes C. Conventional and advanced composites in aerospace industry: Technologies revisited. American Journal of Aerospace Engineering. 2018; **5**(1):9-15. DOI: 10.11648/j.ajae.20180501.12

[76] Klein T, Schachermayer M, Mendez-Martin F, Schöberl T, Rashkova B, Clemens H, et al. Carbon distribution in multi-phase γ-TiAl based alloys and its influence on mechanical properties and phase formation. Acta Materialia. 2015;**94**:205-213. DOI: 10.1016/j.actamat.2015.04.055

[77] Kenel C, Dasargyri G, Bauer T, Colella A, Spierings AB, Leinenbach C, et al. Selective laser melting of an oxide dispersion strengthened (ODS) γ-TiAl alloy towards production of complex structures. Materials & Design. 2017; **134**:81-90. DOI: 10.1016/j.matdes.2017.08.034

[78] Kablov EN, Nochovnaya NA, Panin PV, Alekseev EB, Novak AV. Study of the structure and properties of heat-resistant alloys based on titanium aluminides with gadolinium microadditives. Inorganic Materials: Applied Research. 2017;**8**(4):634-641. DOI: 10.1134/S2075113317040116

[79] Kartavykh AV, Asnis EA, Piskun NV, Statkevich II, Gorshenkov MV, Korotitskiy AV. Room-temperature tensile properties of float-zone processed β-stabilized γ-TiAl (Nb, Cr, Zr) intermetallic. Materials Letters. 2017;**188**:88-91. DOI: 10.1016/j.matlet.2016.10.103

[80] Sienkiewicz J, Kuroda S, Murakami H, Araki H, Giżyński M, Kurzydłowski KJ. Fabrication and oxidation resistance

of TiAl matrix coatings reinforced with silicide precipitates produced by heat treatment of warm sprayed coatings. Journal of Thermal Spray Technology. 2018;**27**(7):1165-1176. DOI: 10.1007/s11666-018-0751-x

[81] Juechter V, Franke MM, Merenda T, Stich A, Körner C, Singer RF. Additive manufacturing of Ti-45Al-4Nb-C by selective electron beam melting for automotive applications. Additive Manufacturing. 2018;**22**:118-126. DOI: 10.1016/j.addma.2018.05.008

[82] Wenjun G, Chao G, Feng L. Microstructures of components synthesized via electron beam selective melting using blended pre-alloyed powders of Ti6Al4V and Ti45Al7Nb. Rare Metal Materials and Engineering. 2015;**44**(11):2623-2627. DOI: 10.1016/S1875-5372(16)60006-1

[83] Liu H, Zhang X, Jiang Y, Zhou R. Microstructure and high temperature oxidation resistance of in-situ synthesized TiN/Ti3Al intermetallic composite coatings on Ti6Al4V alloy by laser cladding process. Journal of Alloys and Compounds. 2016;**670**:268-274. DOI: 10.1016/j.jallcom.2015.10.168

[84] Li W, Li M, Yang Y, Wei Q, Cai D, Liu J, et al. Enhanced compressive strength and tailored microstructure of selective laser melted Ti-46.5 Al-2.5 Cr-2Nb-0.5 Y alloy with different boron addition. Materials Science and Engineering A. 2018;**731**:209-219. DOI: 10.1016/j.msea.2018.06.042

[85] Bresler J, Neumeier S, Ziener M, Pyczak F, Göken M. The influence of niobium, tantalum and zirconium on the microstructure and creep strength of fully lamellar γ/α2 titanium aluminides. Materials Science and Engineering A. 2019;**744**:46-53. DOI: 10.1016/j.msea.2018.11.152

[86] Jiang H, Tian S, Guo W, Zhang G, Zeng S. Hot deformation behavior and deformation mechanism of two TiAl-Mo alloys during hot compression. Materials Science and Engineering A. 2018;**719**:104-111. DOI: 10.1016/j.msea.2018.01.133

[87] Carrullo JC, Falcón JC, Borrás VA. Influence of process parameters and initial microstructure on the oxidation resistance of Ti48Al2Cr2Nb coating obtained by laser metal deposition. Surface and Coatings Technology. 2019;**358**:114-124. DOI: 10.1016/j.surfcoat.2018.11.015

[88] Dai J, Zhang F, Wang A, Yu H, Chen C. Microstructure and properties of Ti-Al coating and Ti-Al-Si system coatings on Ti-6Al-4V fabricated by laser surface alloying. Surface and Coatings Technology. 2017;**309**:805-813. DOI: 10.1016/j.surfcoat.2016.10.082

[89] Maliutina IN, Si-Mohand H, Sijobert J, Bertrand P, Lazurenko DV, Bataev IA. Structure and oxidation behavior of γ-TiAl coating produced by laser cladding on titanium alloy. Surface and Coatings Technology. 2017;**319**:136-144. DOI: 10.1016/j.surfcoat.2017.04.008

[90] Zhang Y, Bandyopadhyay A. Direct fabrication of compositionally graded Ti-Al2O3 multi-material structures using laser engineered net shaping. Additive Manufacturing. 2018;**21**:104-111. DOI: 10.1016/j.addma.2018.03.001

[91] Majumdar JD, Rittinghaus SK, Wissenbach K, Höche D, Blawert C, Weisheit A. Direct laser cladding of the silicide dispersed titanium aluminide (Ti45Al5Nb0. 5Si) composites. Optics & Laser Technology. 2018;**106**:182-190. DOI: 10.1016/j.optlastec.2018.03.012

[92] Guangyu Y, Wenpeng J, Pei Z, Liang J, Nan L, Jian W, et al. Microstructures of as-fabricated and post heat treated Ti-47Al-2Nb-2Cr alloy produced by selective Electron beam melting (SEBM). Rare Metal Materials and Engineering. 2016;**45**(7):1683-1686. DOI: 10.1016/S1875-5372(16)30140-0

[93] Klein T, Holec D, Clemens H, Mayer S. Pathways of phase transformation in β-phase-stabilized σ/γ-TiAl alloys subjected to two-step heat treatments. Scripta Materialia. 2018;**149**:70-74. DOI: 10.1016/j.scriptamat.2018.02.009

[94] Qiang F, Kou H, Tang B, Song L, Li J. Effect of cooling rate on microstructure evolution of Ti-45Al-8.5 Nb-0.2 W-0.2 B-0.02 Y alloy during multi-step heat treatment. Materials Characterization. 2018;**145**:210-217. DOI: 10.1016/j.matchar.2018.08.031

[95] Rittinghaus SK, Hecht U, Werner V, Weisheit A. Heat treatment of laser metal deposited TiAl TNM alloy. Intermetallics. 2018;**95**:94-101. DOI: 10.1016/j.intermet.2018.02.002

[96] Li W, Liu J, Zhou Y, Wen S, Tan J, Li S, et al. Texture evolution, phase transformation mechanism and nanohardness of selective laser melted Ti-45Al-2Cr-5Nb alloy during multi-step heat treatment process. Intermetallics. 2017;**85**:130-138. DOI: 10.1016/j.intermet.2017.01.016

[97] Tebaldo V, Faga MG. Influence of the heat treatment on the microstructure and machinability of titanium aluminides produced by electron beam melting. Journal of Materials Processing Technology. 2017; **244**:289-303. DOI: 10.1016/j.jmatprotec.2017.01.037

[98] Zhao ET, Niu HZ, Zhang SZ, Feng L, Yang SY. Microstructural control and mechanical properties of a β-solidified γ-TiAl alloy Ti-46Al-2Nb-1.5V-1Mo-Y. Materials Science and Engineering A. 2017;**701**:1-6. DOI: 10.1016/j.msea.2017.06.065

[99] Fan J, Liu J, Tian S, Wu S, Wang S, Gao H, et al. Effect of solidification parameters on microstructural characteristics and mechanical properties of directionally solidified binary TiAl alloy. Journal of Alloys and Compounds. 2015;**650**:8-14. DOI: 10.1016/j.jallcom.2015.05.160

[100] Majumdar T, Eisenstein N, Frith JE, Cox SC, Birbilis N. Additive manufacturing of titanium alloys for Orthopedic applications: A materials science viewpoint. Advanced Engineering Materials. 2018;**20**: 1800172. DOI: 10.1002/adem.201800172

Chapter 11

Study of the Wind Speed Forecasting Applying Computational Intelligence

Pedro Junior Zucatelli, Erick Giovani Sperandio Nascimento, Alex Álisson Bandeira Santos, Alejandro Mauricio Gutiérrez Arce and Davidson Martins Moreira

Abstract

The conventional sources of energy such as oil, natural gas, coal, or nuclear are finite and generate environmental pollution. Alternatively, renewable energy source like wind is clean and abundantly available in nature. Wind power has a huge potential of becoming a major source of renewable energy for this modern world. It is a clean, emission-free power generation technology. Wind energy has been experiencing very rapid growth in Brazil and in Uruguay; therefore, it's a promising industry in these countries. Thus, this rapid expansion can bring several regional benefits and contribute to sustainable development, especially in places with low economic development. Therefore, the scope of this chapter is to estimate short-term wind speed forecasting applying computational intelligence, by recurrent neural networks (RNN), using anemometers data collected by an anemometric tower at a height of 100.0 m in Brazil (tropical region) and 101.8 m in Uruguay (subtropical region), both Latin American countries. The results of this study are compared with wind speed prediction results from the literature. In one of the cases investigated, this study proved to be more appropriate when analyzing evaluation metrics (error and regression) of the prediction results obtained by the proposed model.

Keywords: atmospheric science, computer science, energy, wind engineering

1. Introduction

Since the Industrial Revolution in the eighteenth century, fossil fuels have been used as an energy source, contributing to increase the concentration of CO_2 (carbon dioxide) in the atmosphere [1]. An increasing global concentration of CO_2 in the atmosphere, from 290.0 parts per million (ppm) in 1870 [2] to 414.0 ppm in 2019 [3], occurred during the period which was marked by the Second and Third Industrial Revolution. This period is characterized by a significant increase in the use of fossil fuels as an energy source. The increased concentration of CO_2 in the atmosphere results in temperature rise. The increase in the temperature is the major cause for all other changes on the earth's climate. The rise in temperatures is causing warming of oceans, melting of ice mass, and increase in evaporation. Due to this increase in CO_2 emissions and its consequences, the traditional concept of global development incorporated the environmental development. This incorporation resulted a broader concept referred to as sustainable development, which is based on the inseparability of economic, social, and environmental development [1]. Therefore, nowadays, integrated renewable energy system-based power generation has enormous growth and enhanced technological development due to increasing worldwide electricity demand, environmental concerns, and financial aspects [4].

In this context, renewable energy is at the center of the transition to a less carbon-intensive and more sustainable energy system. Renewable energy has grown rapidly in recent years, accompanied by sharp cost reductions for solar photovoltaics and wind power in particular [5]. Wind energy, a sustainable and a domestic source of energy that can reduce our dependency on fossil fuels, has developed rapidly in recent years. It's mature technology and comparatively low cost make it promising as an important energy source in the next decades [6]. The electricity sector remains the brightest spot for renewables energy with the exponential growth of wind power in recent years in the world [7]. **Figure 1** shows the global cumulative installed wind capacity 2001–2017 (adapted from [8]).

Brazil is a large country with regard to its sizable power system and continental distances, both in terms of grid extension and generating capacity. A prominent feature of its power system is the significance of its hydropower [9], which accounts for 59.90% [10] of the generation in the interconnected system. By the end of 2018, there were a total of 583 plants/wind farms and 14.71 GW of installed capacity, a 15.19% growth compared to December 2017, when the installed capacity was 12.77 GW. With an additional 1.94 GW, wind power now makes up 9.0% of the nation's power matrix, which also shows the percent contribution from all sources of energy to the electric power grid at the end of 2018. It is important to remember that at the end of 2017, wind power accounted for 8.10% of the energy generated [10].

Uruguay surprisingly obtains 94.0% of its electricity from renewable sources [11]. In addition to old hydropower plants, large investments in solar, wind, and biomass have increased the proportion of these sources to 55.0% of the total energy (see that the global average is 12.0% and the European average is approximately 20.0%). In this way, wind power has attracted attention, and various wind farms

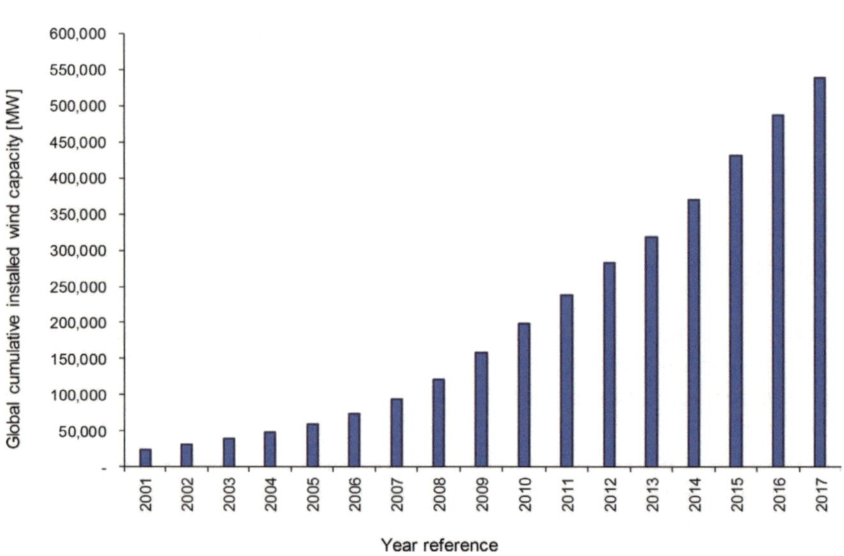

Figure 1.
Wind power global capacity 2001–2017 (adapted from [8]).

have been constructed in Uruguay to harness wind energy. Among the countries of the world, Uruguay ranks fourth in the generation of wind energy, in accordance with [12]. Additionally, Uruguay and Brazil have good relationships, which contribute to its excellent growth with regard to wind and solar energy [13].

Regarding wind power, the variability of wind speed and wind direction throughout the day makes it difficult to decide whether to drive wind turbines, because wind exhibits temporal variations of several orders of magnitude, e.g., short-duration variations (bursts), hourly variations (owing to land and sea breezes), daily variations (owing to the local microclimate), seasonal variations, and annual variations (owing to climatic changes) [13]. The spatial variation of wind energy is also very large. The soil roughness and topography significantly influence the distribution and velocity of winds. Large fluctuations in wind speed make forecasting the power generated by wind turbines difficult; not to mention that economic losses occur if these turbines are subjected to unfavorable weather conditions [14].

Consequently, it is necessary to develop reliable tools to wind speed forecasting, even in the short-range. The interest in applications of mathematical modeling and numerical simulation of the atmosphere for the estimation of wind potential is increasing and driving a significant market. The use of computational models can help both the identification of locations with high wind potential and, when used operationally in daily integrations, in the short-term energy generation forecast [15]. The mainstream models used by scientific researchers can be divided into several categories [13, 16]: physical forecasting models, conventional statistical forecasting models, artificial intelligence forecasting models, statistical machine learning models, fuzzy logic-based models, spatial-correlation forecasting models, and hybrid models.

Computational models can be useful for the identification of locations with high wind potential and, when used operationally in daily integrations, short-term energy generation forecasting [13, 15]. In [13, 17, 18], among others, they obtained good results with small error via mathematical modeling and numerical simulation for short-term prediction using computational intelligence techniques, especially multilayer perceptron neural networks with feed-forward and back-propagation training algorithms. Although the previously cited authors demonstrated the applicability of artificial neural networks (ANN) in the next-step prediction of wind speed, none of them compared the performance of the results of wind speed forecasting 6 h ahead between Colonia Eulacio, Soriano Department, Uruguay (humid subtropical climate region), and Mucuri city, Bahia, Brazil (humid tropical climate region), using meteorological data collected by anemometers and not climatic data from global circulation models shown in [19].

This chapter presents two case studies about short-term wind speed forecasting in Brazil and Uruguay. The chapter is organized as follows. In part 2, the air and the wind power are briefly described. Part 3 describes the materials and methods, computational intelligence, and nowcasting. In part 4, the case studies are proposed. Conclusions are proposed in part 5.

2. The air and wind power

The air in motion—what we commonly call wind—is invisible, yet we see evidence of it nearly everywhere we look. It transports heat, moisture, dust, insects, bacteria, and pollen from one area to another [20]. Inserted in this context, [21] explain that the winds are generated by pressure differences that arise because of unequal heating of the earth's surface. The earth's winds blow in an unending attempt to balance these surface temperature differences. As the zone of maximum solar heating migrates with the seasons—moving northward during the Northern Hemisphere summer and southward as winter approaches—the wind patterns that make up the general circulation also migrate latitudinally. Airflow (or wind) can be divided into three broad categories: waves, turbulence, and mean wind. Each can exist in the boundary layer, where transport of quantities such as moisture, heat, momentum, and pollutants is dominated in the vertical by turbulence and horizontal by the mean wind [22]. Each can exist in the presence of any of the others or separately.

The earth's highly integrated wind system can be thought of as a series of deep rivers of air that encircle the planet. Embedded in the main currents are vortices of various sizes, including hurricanes, tornadoes, and midlatitude cyclones. Like eddies in a stream, these rotating wind systems develop and die out with somewhat predictable regularity. In general, the smallest eddies, such as dust devils, last only a few minutes, whereas larger and more complex systems, such as midlatitude cyclones and hurricanes, may survive for several days [21]. The scales of atmospheric motion shown in **Table 1** illustrate the three major categories of atmospheric circulation: microscale, mesoscale, and macroscale (synoptic scale and global scale).

In short, wind is the movement of air from an area of high pressure to an area of low pressure. In fact, wind exists because the sun unevenly heats the surface of the earth. As hot air rises, cooler air moves in to fill the void. As long as the sun shines, the wind will blow. And wind has long served as a power source to humans [23]. Wind spins the blades, which turn a shaft connected to a generator that produces electricity, in other words, wind turbines convert kinetic energy contained in the wind first into mechanical and then into electrical energy [24]. Wind is a clean source of renewable energy that produces no air or water pollution. And since the wind is free, operational costs are nearly zero once a turbine is erected. Mass production and technology advances are making turbines cheaper, and many governments offer tax incentives to spur wind-energy development [23].

Nowadays, wind turbine technology is considered matured, and the costs of wind energy are low [6]. Industry experts predict that if this pace of growth continues, by 2050 one third of the world's electricity needs will be fulfilled by wind power [23]. Though wind power has performed well in recent years, it also creates a strong environmental impact, such as visual impact, climatic impact, and noise. Although these impacts seem minor when compared with nonrenewable energy, its effect on humans should not be overlooked, due to its potential great development in usage. In short, with proper and supportive policies toward wind power and a good understanding of its environmental impact, wind energy can be a clean and sustainable source of energy that can successfully replace fossil fuels [6].

Scale	Typical size	Phenomenon	Life span
Microscale	0–1.0 km	Small turbulent eddies, thunderstorms	Seconds to minutes
Mesoscale	1.0–100 km	Tornadoes, waterspouts, dust devils, land/sea breeze, mountain/valley breeze	Minutes to hours/days
Synoptic scale	100–5000 km	Hurricanes, tropical storms	Days to weeks
Global scale	1000–40,000 km	Longwaves in the westerlies, trade wind	Weeks to years

Table 1.
Scales of atmospheric motion (adapted from [20, 21]).

3. Materials and methods

In this chapter, we use computational intelligence by artificial neural networks for the next-step prediction of one climatic variable: wind speed. ANN was trained to perform the forecasting of 1 h ahead, and then, using it, the trained network was applied to recursively infer the forecasting for the next 6 h of the wind speed (nowcasting), following the methodology explained in [13]. The activation functions that define the outputs of the neurons in terms of their activity levels, inserted in this simulation, were the sigmoidal function in the form of the hyperbolic tangent function (characterized as continuous, increasing, differentiable, and nonlinear) for hidden layers and linear function to the output layer.

To train the RNN and validate the technique, anemometer data (average hourly values of wind speed, wind direction, and temperature) for 1 year (August 08, 2014, and August 07, 2015) are collected by one tower with anemometer installed at height of 101.8 m to Colonia Eulacio (Uruguay), and data (average hourly values of wind speed, wind direction, temperature, humidity, and pressure) for 1 month (November 30, 2015, until December 31, 2015) are collected by one tower with anemometers installed at height of 100.0 m to Mucuri (Brazil), using the same criteria as in [13], namely, 70% for training/validating data and 30% for simulation. The reason for choosing these periods is these are the months with the totality of data available for the realization of this study.

The Mucuri city (Bahia, Brazil) is located at an altitude of 7.0 m in relation to the sea level, and it has a territorial area of 1775 km^2, approximately. Mucuri's anemometer tower is located in a coastal plain, at a distance of 340.0 m from the sea, with latitude 18°1′31.52″ S and longitude 39°30′51.69″ W (**Figure 2**).

As for Colonia Eulacio Tower in Uruguay, according to datum WGS84, it is located at 33°16′ S, 57°31′ W [25]. The altitude of the installation location (see **Figure 3**) is approximately 100.0 m, and the location is surrounded by fields with plains; thus, it is characterized by noncomplex terrain. The station is owned by the Administración Nacional de Usinas y Transmissiones Eléctricas (UTE), which is a state-owned company in Uruguay that is responsible for the generation, distribution, and commercialization of electrical energy in the country, as cited in [13].

Figure 2.
Location of the Mucuri Tower in Bahia, Brazil.

Figure 3.
Location of the Colonia Eulacio Tower in Soriano Department, Uruguay [21].

The insertion of meteorological parameters as input data contributes to efficient training of the ANN. Seven different ANN configurations are applied for each site and height; then, a quantitative analysis is conducted, and the statistical results (MAE, MSE, and RMSE) are evaluated to select the configuration that best predicts the real data. The proposed ANN configurations to be analyzed are the following ones. For Mucuri (Brazil) the best ANN configuration was Configuration 1 and for Colonia Eulacio (Uruguay) was the Configuration 4.

1. Configuration 1: three layers, nine input nodes (site, Brazil) or seven input nodes (site, Uruguay), nine hidden neurons, and one output node

2. Configuration 2: three layers, nine input nodes (site, Brazil) or seven input nodes (site, Uruguay), six hidden neurons, and one output node

3. Configuration 3: three layers, nine input nodes (site, Brazil) or seven input nodes (site, Uruguay), three hidden neurons, and one output node

4. Configuration 4: three layers, nine input nodes (site, Brazil) or seven input nodes (site, Uruguay), one hidden neuron, and one output node

5. Configuration 5: four layers, nine input nodes (site, Brazil) or seven input nodes (site, Uruguay), nine hidden neurons (first hidden layer) and six hidden neurons (second hidden layer), and one output node

6. Configuration 6: four layers, nine input nodes (site, Brazil) or seven input nodes (site, Uruguay), six hidden neurons (first hidden layer) and three hidden neurons (second hidden layer), and one output node

7. Configuration 7: four layers, nine input nodes (site, Brazil) or seven input nodes (site, Uruguay), one hidden neuron (first hidden layer) and one hidden neuron (second hidden layer), and one output node

For statistical analysis of wind speed prediction results at the above sites, the following statistical indicators were applied: Pearson's correlation coefficient (r or Pearson's r), coefficient of determination (R^2 or R-squared), mean absolute percentage error (MAPE), mean absolute error (MAE), mean squared error (MSE), and root mean square error (RMSE). Pearson's correlation coefficient ranges from −1.0 to 1.0. Values close to 0.0 are adequate for the MAE, MSE, and RMSE, values close to 0.0% are adequate for the MAPE, and values close to 1.0 are adequate for the R-squared. The software used to program and perform this computational procedure was MATLAB version 7.10.0 (2010) (personal computer, 8 GB RAM), as the methodology applied in [13].

3.1 Computational intelligence and nowcasting

Computational intelligence (CI) is the theory, design, application, and development of biologically and linguistically motivated computational paradigms. Over the last few years, there has been an explosion of research on machine learning and deep learning. Nowadays, deep learning has become the core method for artificial intelligence (AI) [26]. AI is one of the newest fields in science and engineering. AI currently encompasses a huge variety of subfields, ranging from the general (learning and perception) to the specific, such as playing chess, proving mathematical theorems, writing poetry, driving a car on a crowded street, diagnosing diseases, and predicting the conditions of the atmosphere for a given location and time. AI is relevant to any intellectual task; it is truly a universal field [27]. In fact, some of the most successful AI systems are based on CI.

In the early days of artificial intelligence, the field rapidly tackled and solved problems that are intellectually difficult for human beings but relatively straightforward for computers—problems that can be described by a list of formal, mathematical rules [28]. Many artificial intelligence tasks can be solved by designing the right set of features to extract for that task, then providing these features to a simple machine learning algorithm. The most widely used artificial neuron model is the perceptron proposed in [29, 30]. This model defines a neuron composed of inputs, a summation and an activation function. The value of each input is multiplied by a weight, and the weighted values of the inputs are summed to yield the result of the sum which is used as the input of the activation function. To teach (train) the neuron, the weights are modified so that the output obtained corresponds to the desired value [30].

The multilayer perceptron (MLP) consists of a system of simple interconnected neurons, or nodes, which is a model representing a nonlinear mapping between an input vector and an output vector. The architecture of a MLP is variable but in general will consist of several layers of neurons. The input layer plays no computational role but merely serves to pass the input vector to the network. The terms input and output vectors refer to the inputs and outputs of the MLP and can be represented as single vectors [31]. Moreover, in relation to recurrent neural network (RNN), the definition is that they are powerful sequence-processing models that are equipped with a memory from recurrent feedback connections. One of the current main challenges of RNN is to dynamically adapt to multiple temporal resolutions and scales in order to learn hierarchical representations in time. Since they operate in discrete time steps and update at every time step, it is generally difficult to learn temporal features that have a significantly different resolution than their input frequency [32].

Predicting the short-term power output of a wind turbine (wind energy converter) is an important task for the efficient management of smart grids. Short-term forecasting at the minute scale also is known as nowcasting. By definition nowcasting refers to short lead time weather forecasts, the US National Weather Service specifies zero to 3 h, though forecasts up to 6 h may be called nowcasts by some agencies [33]. Nowcasting is critical when managing operations of the smart grid, such as system integration, ensuring power continuity and managing ramp rates [34]. In this chapter, nowcasting refers to short-term wind speed forecasting 6 h ahead. In [35] they described and evaluated a proposal for nowcasting wind speed for wind farm locations from historical time series, based on the method of regression by support vectors (in short, nowcasting of wind speed using support vector regression: experiments with time series from Gran Canaria).

4. Brazilian and Uruguayan case study: Mucuri municipality, tropical region, and Colonia Eulacio, subtropical region

The figures below show the time series used in the models which consist of 744 h for Mucuri in Brazil (**Figure 4**) and 8760 h for Colonia Eulacio in Uruguay (**Figure 5**), corresponding to hourly mean wind speed data. As can be observed in the figures, there is noticeable data randomness, and it is difficult to find a series tendency or seasonality.

In this sense, a descriptive statistic regarding wind speed at different sites is shown in **Table 2**. It can be noted that wind speed data measured at Colonia Eulacio has a higher variability.

Tables 3–6 show the evaluation metrics of the prediction results obtained by the proposed model. **Table 3** presents simulation results referring to errors in wind speed forecasting 1, 3, and 6 h ahead for Mucuri using the RNN model. In **Table 4**, we present the results for regression for Mucuri. In **Table 5**, we can see errors in wind speed forecasting 1, 3, and 6 h ahead for Colonia Eulacio using the RNN model. **Table 6** shows the regression results to Colonia Eulacio. The percentage of the data of a factor of two is a fraction of data [%] for 0.5 ≤ wind predicted/wind anemometer ≤ 2.0. **Table 4** shows that the percentage of the data of a factor of two [%] to Mucuri is bigger than the percentage of the data of a factor of two [%] to Colonia Eulacio (see **Table 6**).

Figures 6–8 show the results of six-step predictions of the wind speed series to Colonia Eulacio and to Mucuri. In **Figure 6**, we observe wind speed forecasting results of the model with RNN in six-step ahead for Colonia Eulacio, Uruguay. **Figure** 7 presents wind speed forecasting 6 h ahead in a period of 744 ho of

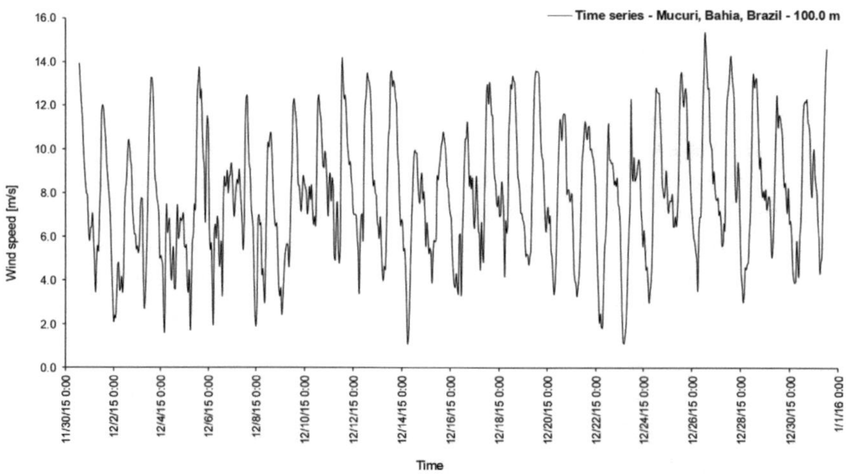

Figure 4.
The experimental wind speed time series—Mucuri (Brazil).

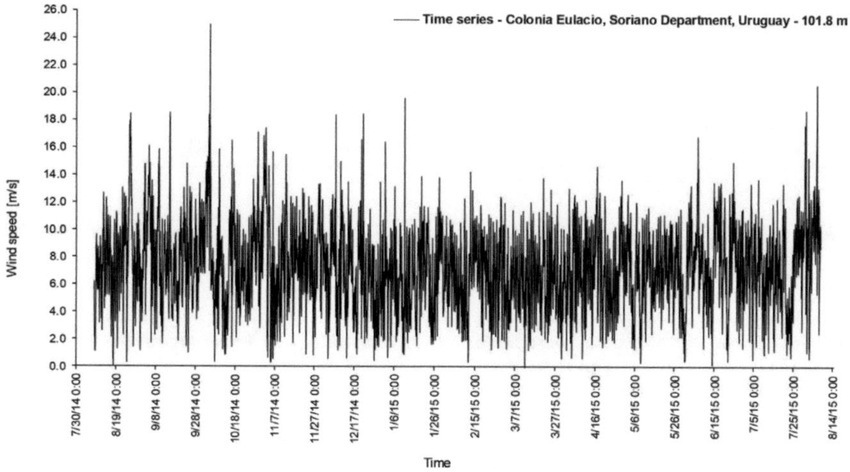

Figure 5.
The experimental wind speed time series—Colonia Eulacio (Uruguay).

Site	Arithmetic mean of wind speed [m/s]	Variance [m^2/s^2]	Standard deviation [m/s]
Mucuri	7.91	8.53	2.92
Colonia Eulacio	7.21	9.02	3.00

Table 2.
Statistics (Mucuri and Colonia Eulacio).

Prediction horizon [h]	MAE	MSE	RMSE	MAPE [%]
1	0.839	1.111	1.054	11.07
3	1.385	3.154	1.775	17.63
6	1.779	5.108	2.260	21.26

Table 3.
Simulation results (errors): wind speed forecasting 1, 3, and 6 h ahead for Mucuri using the RNN model.

Prediction horizon [h]	Pearson correlation coefficient	Coefficient R^2	Percentage of the data of a factor of two [%]
1	0.940	0.885	99.48
3	0.850	0.723	98.95
6	0.742	0.550	98.94

Table 4.
Simulation results (regression): wind speed forecasting 1, 3, and 6 h ahead for Mucuri using the RNN model.

anemometric tower measurements. In **Figure 8**, we can see wind speed forecasting results of the model with RNN in six-step ahead for Mucuri, Brazil.

Figures 9 and **10** show the multistep root mean square error (RMSE) evaluation of RNN and MLP for Mucuri, Brazil, and Colonia Eulacio, Uruguay. It is observed in **Figure 9** that as the prediction horizon increases, RNN were more efficient when compared to those employed in the study which applies MLP referenced by [36]. These results indicate that if we want a higher accuracy in the result to Mucuri, we must use a

Prediction horizon [h]	MAE	MSE	RMSE	MAPE [%]
1	0.895	1.408	1.187	16.08
3	1.687	4.710	2.170	30.65
6	2.266	8.051	2.837	39.86

Table 5.
Simulation results (errors): wind speed forecasting 1, 3, and 6 h ahead for Colonia Eulacio using the RNN model.

Prediction horizon [h]	Pearson correlation coefficient	Coefficient R^2	Percentage of the data of a factor of two [%]
1	0.922	0.849	98.44
3	0.729	0.531	93.60
6	0.543	0.295	88.52

Table 6.
Simulation results (regression): wind speed forecasting 1, 3, and 6 h ahead for Colonia Eulacio using the RNN model.

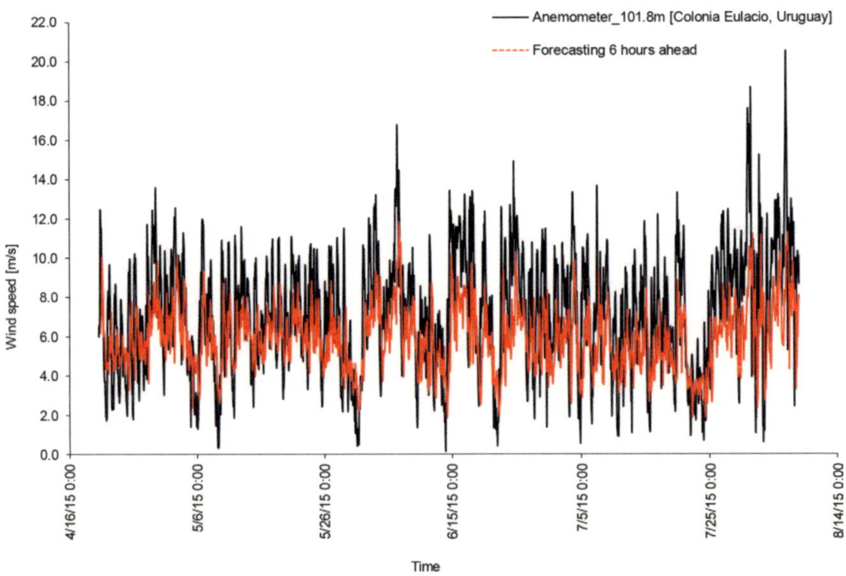

Figure 6.
Wind speed forecasting results of the model with RNN in six-step ahead (Colonia Eulacio, Uruguay).

recurrent neural network. A recurrent neural network allows self-loops and backward connections between all neurons in the network. That enables the networks to do temporal processing and learn sequences, e.g., temporal association/prediction.

Unlike the previous comparison, for Colonia Eulacio's anemometers data, the study which applies MLP referenced by [13] was more appropriate (**Figure 10**) than this study that applies RNN.

The multistep Pearson correlation coefficient evaluation of different architecture and site is shown in **Figure 11**.

The computational cost employed to simulate wind speed prediction through RNN for Colonia Eulacio is not viable when compared to the application of MLP. In

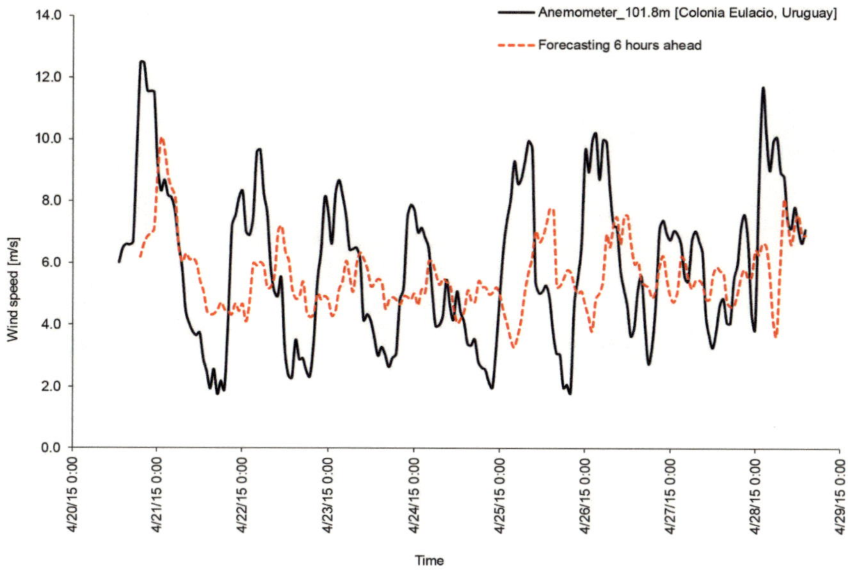

Figure 7.
Wind speed forecasting results of the model with RNN in six-step ahead (Colonia Eulacio, Uruguay), April 20, 2015, at 13:00 to April 28, 2015, at 14:00.

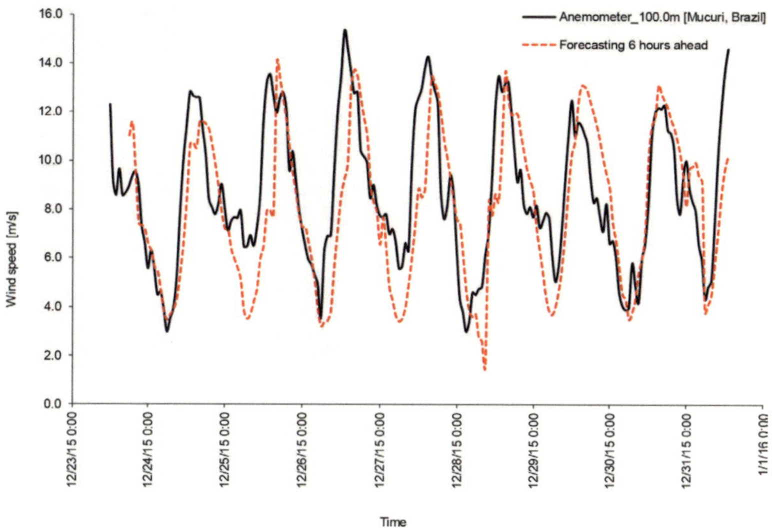

Figure 8.
Wind speed forecasting results of the model with RNN in six-step ahead (Mucuri, Brazil).

contrast, for Mucuri, it is considerably more viable to apply RNN, as can be seen from the figure above. Lastly, nowadays, adopting renewable energy has become a national energy policy for many countries due to concerns with pollution from fossil fuel consumption and climate change. Regarding wind energy, the accurate prediction of wind is crucial in managing the power load. Thus, this work presented the short-term wind speed forecasting for two representative sites in South America, Brazil, and Uruguay, which are the most important countries in terms of renewable energy production in Latin America.

Each scientific study on wind speed prediction has its own characteristics, such as the height of the anemometer that records atmospheric data (e.g., wind speed,

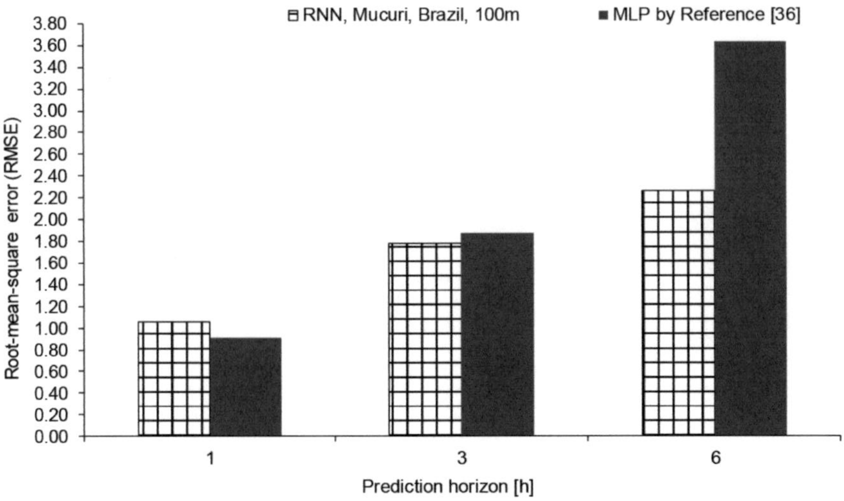

Figure 9.
The multistep root mean square error evaluation of different architecture for Mucuri, Brazil.

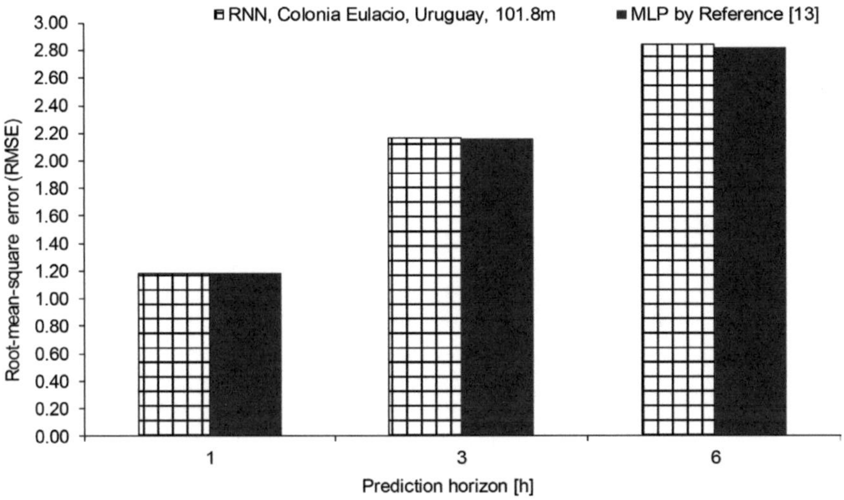

Figure 10.
The multistep root mean square error evaluation of different architecture for Colonia Eulacio, Uruguay.

direction, temperature, humidity, and atmospheric pressure) and the time series of this atmospheric data. These data are applied to train and test the efficiency of the ANN. On the accuracy of the use of ANN in the estimation of short-term wind speed and wind power forecasting, we can mention these earlier studies (see **Table 7**). The MAE average value for 1 h ahead was 0.847 m/s and for 3 h ahead was 1.420 m/s.

Other scientific research on wind speed has been developed. We can cite the works in [39, 40]. In [39] they analyzed the time series of wind speeds in Mucuri, Mucugê, and Esplanada, cities of the state of Bahia, and the Abrolhos Archipelago, Brazil, through the use of the detrended fluctuation analysis technique to verify the existence of long-range correlations and associated power laws. Already [40] describes a short-term wind energy forecasting tool based on a run set forecasting system of the WRF-GFS model that has been operationally implemented in the electricity system in Uruguay with estimates for Brazil wind energy production.

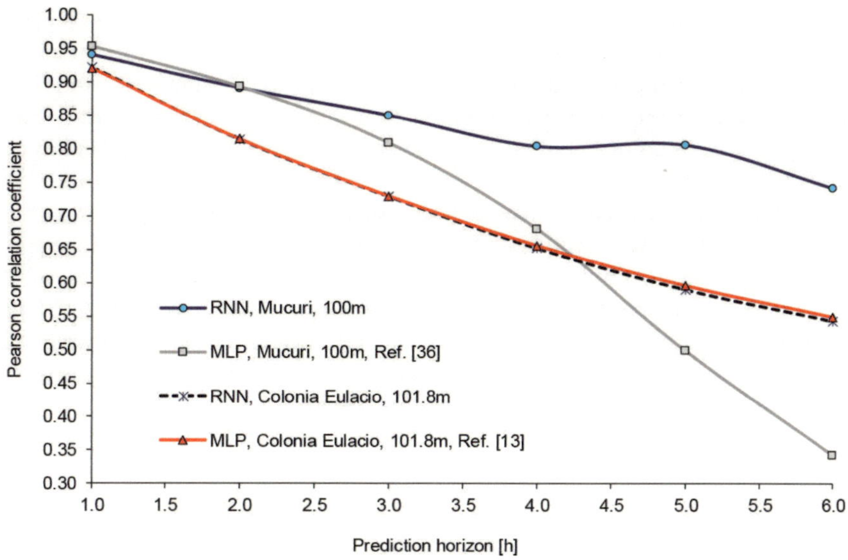

Figure 11.
The multistep Pearson correlation coefficient evaluation of different architecture and site.

Prediction horizon [h]	Mucuri: this study	Colonia Eulacio: this study	Reference [13]	Reference [36]	Reference [37]	Reference [38]
			Mean absolute error			
1	0.839	0.895	0.892	0.720	1.050	0.684
3	1.385	1.687	1.678	1.370	1.230	1.168

Table 7.
Accuracy of the use of ANN in the estimation of wind speed and wind power.

5. Conclusions

The present chapter aimed to define the most efficient RNN configuration to predict the wind speed for 1 h and, after that, to apply it for 6 h ahead, using as reference observational data collected from two anemometric towers, with anemometers installed at 100.0 and 101.8 m height, located, respectively, in a tropical region in Mucuri, Bahia, northeastern Brazil, and in a subtropical region in Colonia Eulacio, Soriano Department, Uruguay. It has been shown graphically and verified through numerical simulations that the RNN was better than MLP in Mucuri and worst in Colonia Eulacio.

In the light of the statistical results recorded in this work, the application of computational intelligence is a viable alternative for the predictability of wind speed and, in this way, wind power generation, mainly due to the low computational cost; however, one must choose the ANN architecture that best suits the project, as well as quantitatively and qualitatively analyzes the available data that will feed the network, since these variables directly impact the results of the forecast.

The results of the short-term wind speed forecasting showed good accuracy at all the anemometer heights tested. Therefore, the proposed short-term wind speed forecasting method is an important scientific contribution for reliable large-scale wind power forecasting and integration in tropical and subtropical regions, like in Brazil and Uruguay.

The suggestion for improving the accuracy of ANN for higher lead time is wavelet packet decomposition because the empirical wavelet transform can effectively identify and extract a finite number of intrinsic modes of a wind speed time series and thus improving the accuracy of the supervised machine learning; other suggestion is to apply the wind speed x-axis component and wind speed y-axis component ANN's input.

We can suggest as future work to use the Mucuri, Colonia Eulacio, and other observational data collected in different heights in Brazil and Uruguay to perform the prediction of the wind speed more accurately in the short-term and in the medium-term using computational intelligence by long short-term memory (LSTM) and gated recurrent unit (GRU) and to compare these results with the output produced by numerical and meteorological modeling using the weather research and forecasting (WRF) model, for example. Wind ramp and greater forecasting horizons are also a great subject of research.

Acknowledgements

We thank SENAI CIMATEC and FAPESB for their computational and financial support, as well as UTE and UFES.

Conflict of interest

The authors declare no conflict of interest.

Thanks

We thank IntechOpen, the world's leading publisher of Open Access books, for this opportunity.

Author details

Pedro Junior Zucatelli[1*], Erick Giovani Sperandio Nascimento[2], Alex Álisson Bandeira Santos[2], Alejandro Mauricio Gutiérrez Arce[3] and Davidson Martins Moreira[2]

1 Federal University of Espirito Santo—UFES, Vitória, Brazil

2 Manufacturing and Technology Integrated Campus—SENAI CIMATEC, Salvador, Brazil

3 Universidad de la Republica—UDELAR, Montevideo, Uruguay

*Address all correspondence to: pedrojrzucatelli@gmail.com

References

[1] Zucatelli PJ, Meneguelo AP, Chaves GLD, Tosta MCR. The application, required investments and operational costs of geological CO_2 sequestration: A case study. Research, Society and Development. 2019;**8**(6):1-28. DOI: 10.33448/rsd-v8i6.1023. ISSN 2525-3409

[2] UNEP/GRID-Arendal. Global Atmospheric Concentration of CO2. UNEP/GRID-Arendal Maps and Graphics Library. 1999. Available from: http://www.grida.no/resources/5500 [Accessed July, 2019]

[3] NOAA. Global Greenhouse Gas Reference Network. 2019. Available from: https://www.esrl.noaa.gov/gmd/ccgg/trends/graph.html [Accessed July, 2019]

[4] Abarzadeh M, Kojabadi HM, Chang L. Study of novel power electronic converters for small scale wind energy conversion systems. In: Cao W, Hu Y, editors. Renewable Energy—Utilisation and System Integration. Rijeka: IntechOpen; 2016. DOI: 10.5772/62477. Available from: https://www.intechopen.com/books/renewable-energy-utilisation-and-system-integration/study-of-novel-power-electronic-converters-for-small-scale-wind-energy-conversion-systems

[5] IEA. Renewables. 2018. Available from: https://www.iea.org/topics/renewables/ [Accessed July, 2019]

[6] Leung DYC, Yang Y. Wind energy development and its environmental impact: A review. Renewable and Sustainable Energy Reviews. 2012;**16**(1):1031-1039. DOI: 10.1016/j.rser.2011.09.024

[7] Renewables. Global Status Report: 02 Market and Industry Trends. 2017. Available from: http://www.ren21.net/gsr-2017/chapters/chapter_02/chapter_02/#figure_26 [Accessed July, 2019]

[8] GWEC. Global Wind Statistics 2017, GWEC. 2018. Available from: http://gwec.net/wp-content/uploads/vip/GWEC_PRstats2017_EN-003_FINAL.pdf [Accessed July, 2019]

[9] Rego EE, Ribeiro CO. Successful Brazilian experience for promoting wind energy generation. The Electricity Journal. 2018;**31**(2):13-17. DOI: 10.1016/j.tej.2018.02.003

[10] ABEEólica. Annual Wind Energy Report 2018. ABEEólica, the Brazilian Wind Power Association. Available from: http://abeeolica.org.br/dados-externos/ [Accessed July, 2019]

[11] Watts J. Uruguay Makes Dramatic Shift to Nearly 95% Electricity from Clean Energy: Keep it in the Ground Renewable Energy. 2015. Available from: https://www.theguardian.com/environment/2015/dec/03/uruguay-makes-dramatic-shift-to-nearly-95-clean-energy [Accessed July, 2019]

[12] REN21. Renewables 2017 Global Status Report. REN21 Secretariat, Paris. 2017. Available from: http://www.ren21.net/gsr-2017/chapters/chapter_01/chapter_01/ [Accessed July, 2019]

[13] Zucatelli PJ, Nascimento EGS, Aylas GYR, Souza NBP, Kitagawa YKL, Santos AAB, et al. Short-term wind speed forecasting in Uruguay using computational intelligence. Heliyon. 2019;**5**(5):e01664. DOI: 10.1016/j.heliyon.2019.e01664

[14] Esmaeili MA, Twomey J. Self-organizing map (SOM) in wind speed forecasting: A new approach in computational intelligence (CI) forecasting methods. In: Proceedings of the ASME/ISCIE 2012 International Symposium on Flexible Automation,

ISFA; June 18-20, 2012; St. Louis, Missouri, USA. 2012. DOI: 10.1115/ISFA2012-7241

[15] Peng H, Liu F, Yang X. A hybrid strategy of short term wind power prediction. Renewable Energy. 2013;**50**:590-595. DOI: 10.1016/j.renene.2012.07.022

[16] Jiang P, Wang Y, Wang J. Short-term wind speed forecasting using a hybrid model. Energy. 2017;**119**:561-577. DOI: 10.1016/j.energy.2016.10.040

[17] Wasilewskia J, Baczynski D. Short-term electric energy production forecasting at wind power plants in pareto-optimality context. Renewable and Sustainable Energy Reviews. 2017;**69**:177-187. DOI: 10.1016/j.rser.2016.11.026

[18] Alencar DB, Affonso CM, Oliveira RCL, Rodríguez JLM, Leite JC, FilhoJCR.Differentmodelsforforecasting wind power generation: Case study. Energies. 2017;**10**(12):1976. DOI: 10.3390/en10121976

[19] López-Manrique LM, Macias-Melo EV, May Tzuc O, Bassam A, Aguilar-Castro KM, Hernández-Pérez I. Assessment of resource and forecast modeling of wind speed through an evolutionary programming approach for the north of Tehuantepec Isthmus (Cuauhtemotzin, Mexico). Energies. 2018;**11**:3197. DOI: 10.3390/en11113197

[20] Ahrens CD. Meteorology Today: An Introduction to Weather, Climate, and the Environment. In: Ahrens CD, editor. 9th ed. USA: Brooks/Cole, Cengage Learning, Belmont; 2009. 621 p

[21] Lutgens FK, Tarbuck EJ. In: Lutgens FK, Tarbuck EJ, editors. The Atmosphere: An Introduction to Meteorology. 6th ed. New Jersey: Prentice Hall, Inc.; 1995

[22] Stull RB. An Introduction to Boundary Layer Meteorology. Dordrecht, The Netherlands: Kluwer Academic Publishers; 1988. 666 p

[23] National Geographic. Wind Power. Available from: https://www.nationalgeographic.com/environment/global-warming/wind-power/ [Accessed July, 2019]

[24] He W, Ge SS. Vibration control of a nonuniform wind turbine tower via disturbance observer. IEEE/ASME Transactions on Mechatronics. 2015;**20**:237-244. DOI: 10.1109/TMECH.2014.2313876

[25] Lucas EA, Arce AMG, Moraes MR, Boezio GC, Ottieri JC. Statistical description of diurnal cycle of wind profile in the first 100 meters of height of the Planetary Boundary Layer. In: Colonia Eulacio Uruguay. Ciência e Natura v.38 Ed. Especial-IX Brazilian Micrometeorology Workshop. 2016. pp. 426-434. DOI: 10.5902/2179460X20308

[26] IEEE Computational Intelligence Society. Available from: https://cis.ieee.org/about/what-is-ci [Accessed July, 2019]

[27] Russel S, Norvig P. Artificial Intelligence: A Modern Approach. In: Russell SJ, Norvig P, editors. 3rd ed. New Jersey, USA: Pearson Education, Inc.; 2010. 1153 p

[28] Data Science Academy. Available from: http://www.deeplearningbook.org/contents/intro.html [Accessed July, 2019]

[29] Rosenblatt F. The perceptron: A probabilistic model for information storage and organization in the brain. Psychological Review. 1958;**65**(6):386-408. DOI: 10.1037/h0042519

[30] Monteiro RLS, Pereira HBB, Moreira DM. Study of the impact of the topology of artificial neural networks

for the prediction of meteorological data. In: Mendes Platt G et al., editors. eBook of Computational Intelligence, Optimization and Inverse Problems with Applications in Engineering. Cham, Switzerland: Springer; 2019. pp. 201-214. DOI: 10.1007/978-3-319-96433-1_10. ISBN: 978-3-319-96433-1

[31] Gardner MW, Dorling SR. Artificial neural networks (the multilayer perceptron)—A review of applications in the atmospheric sciences. Atmospheric Environment. 1998;**32**(14-15):2627-2636. DOI: 10.1016/S1352-2310(97)00447-0

[32] Alpay T, Abawi F, Wermter S. Preserving activations in recurrent neural networks based on surprisal. Neurocomputing. 2019;**342**:75-82. DOI: 10.1016/j.neucom.2018.11.092

[33] Kuikka I. Wind nowcasting: Optimizing runway in use. In: Technical Report. Systems Analysis Laboratory, Helsinki University of Technology; 2009. Available from: http://salserver.org.aalto.fi/vanhat_sivut/Opinnot/Mat-2.4108/pdf-files/ekui09.pdf. [Accessed: 08-10-19]

[34] Zhang J, Verschae R, Nobuhara S, Lalonde J-F. Deep photovoltaic nowcasting. Solar Energy. 2018;**176**(2018):267-276. DOI: 10.1016/j.solener.2018.10.024

[35] Espino I, Hernández M. Nowcasting of wind speed using support vector regression. Experiments with time series from Gran Canaria. RE&PQJ. 2011;**1**(9):700-705. DOI: 10.24084/repqj09.428

[36] Zucatelli PJ, Nascimento EGS, Aylas GYR, Souza NBP, Kitagawa YKL, Santos AAB, et al. Short-term wind speed forecasting using computational intelligence in Bahia, Brazil. In: Proceedings of the Air Pollution Conference Brazil and 4th CMAS South America, 2019. Extended Abstract. Belo Horizonte. 2019

[37] Ahmed A, Khalid M. An intelligent framework for short-term multi-step wind speed forecasting based on functional networks. Applied Energy. 2018;**225**(2018):902-911. DOI: 10.1016/j.apenergy.2018.04.101

[38] Liu H, Mi X, Li Y. Comparison of two new intelligent wind speed forecasting approaches based on wavelet packet decomposition, complete ensemble empirical mode decomposition with adaptive noise and artificial neural networks. Energy Conversion and Management. 2018;**155**(2018):188-200. DOI: 10.1016/j.enconman.2017.10.085

[39] Santos JVC, Moreira DM, Moret MA, Nascimento EGS. Analysis of long-range correlations of wind speed in different regions of Bahia and the Abrolhos Archipelago, Brazil. Energy. 2019;**167**:2019. DOI: 10.1016/j.energy.2018.11.015

[40] Gutiérrez A, Almeida E, Porrini C, Romero M, Ortelli S, Cazes G, et al. Pronósticos operativos de energía eólica para la gestión de los sistemas eléctricos en la región. Enerlac. 2018;**II**(1):28-41. Available from: http://enerlac.olade.org/index.php/ENERLAC/article/view/57/27. [Accessed September, 2019]